2024 Global Conference on Process Safety and Big Data

AA008980

Frankfurt, Germany
29-30 October 2024

ISBN: 979-8-3313-1053-0

Printed from e-media with permission by:

Curran Associates, Inc.
57 Morehouse Lane
Red Hook, NY 12571

Some format issues inherent in the e-media version may also appear in this print version.

Copyright© (2024) by AIChE
All rights reserved.

Printed with permission by Curran Associates, Inc. (2025)

For permission requests, please contact AIChE
at the address below.

AIChE
120 Wall Street, FL 23
New York, NY 10005-4020

Phone: (800) 242-4363
Fax: (203) 775-5177

www.aiche.org

Additional copies of this publication are available from:

Curran Associates, Inc.
57 Morehouse Lane
Red Hook, NY 12571 USA
Phone: 845-758-0400
Fax: 845-758-2633
Email: curran@proceedings.com
Web: www.proceedings.com

TABLE OF CONTENTS

GENERATIVE ARTIFICIAL INTELLIGENCE AND BIG DATA ANALYTICS FOR PROCESS SAFETY

PROCESS SAFETY MANAGEMENT AND PERMIT TO WORK

Digitalization Journey at Oqgn

Conference

Global Conference on Process Safety and Big Data

Year

2024

Proceeding

2024 Global Conference on Process Safety and Big Data

Group

General Program

Session

2024 Global Conference on Process Safety and Big Data Keynote Presentations

Time

Tuesday, October 29, 2024 - 9:15am to 9:45am

Authors

Alburaidi, M. - Presenter

OQGN, a prominent entity in the oil and gas sector, is Oman's exclusive owner and operator of the natural gas transmission network. Over the last two decades, OQGN has successfully transformed the business from a small gas transport company into a critical infrastructure player with a 4030 km pipeline network connecting producers with consumers in Oman.

Prior to OQGN's digitalization journey, process safety systems were entirely paper-based, with no tracking system or real-time data capabilities, significantly impairing the implementation of the process management system. Reporting of process safety performance indicators was manually handled through email, which compromised report accuracy and led to considerable delays in data receipt. Moreover, the absence of a dashboard and historical data made it difficult to identify improvement areas and support the decision-making process. Due to these significant obstacles, it was crucial for OQGN to embark on a digitalization journey to overcome these challenges and enhance overall efficiency and effectiveness.

The digitalization at OQGN was initiated with extensive benchmarking efforts to identify the best platforms for our digitalization journey. Following this benchmarking, we selected the optimal platforms based on well-defined criteria and collaborated with various systems and vendors to digitalize the existing process safety systems. This involved comprehensive steps from system configuration and user acceptance testing to full system rollout, all carried out with the active engagement of the stakeholders . Additionally, training sessions were conducted for end users to ensure they possessed the necessary skills and competency to effectively utilize the new digital systems, thereby ensuring high-quality implementation and smooth transition.

OQGN has achieved remarkable advancements in its digitalization journey, emphasizing initiatives aimed at enhancing operational efficiency and safety. Key initiatives include the deployment of an electronic Management of Change (MOC) system, an electronic Temporary Repair system, a digital Process Safety Management (PSM) dashboard, and digital Process Safety Performance Indicators (PSPIs) reporting.

Integral to this transformation is the electronic Management of Change (MOC) system, which has replaced manual, paper-based processes. Additionally, the electronic Temporary Repair system has revolutionized the management of temporary repairs by enabling real-time tracking and status updates.

Furthermore, OQGN has developed a digitalized PSM dashboard that consolidates data from diverse sources to provide a comprehensive, real-time perspective on process safety performance throughout the organization. The introduction of digital Process Safety Performance Indicators (PSPIs) reporting automates the collection, analysis, and dissemination of process safety performance indicators, ensuring more precise and timely reporting.

The digitalization efforts at OQGN have resulted in significant benefits, such as cost savings, reduced man-hours, improved system performance monitoring, enhanced decision-making, led to more effective reporting, and contribute substantially to the organization's overall digital transformation.

OQGN's experience underscores the importance of embracing digitalization to drive sustainable growth and innovate business practices. The company's commitment to digital transformation highlights the value of leveraging technology to achieve operational excellence and ensure long-term success in the industry.

Leveraging Artificial Intelligence to Improve Process Safety in the Chemical Industry

Conference

Global Conference on Process Safety and Big Data

Year

2024

Proceeding

2024 Global Conference on Process Safety and Big Data

Group

General Program

Session

2024 Global Conference on Process Safety and Big Data Keynote Presentations

Time

Wednesday, October 30, 2024 - 12:15pm to 2:15pm

Authors

Bohm, P. - Presenter, Accenture
Sousa, R., Accenture

A typical chemical plant has around 5,000 pieces of equipment with dozens of Standard Operating Procedures (SOPs) manuals associated with each one. SOPs are critical for plant safety, troubleshooting, continuous improvement and compliance. However, they require a tremendous amount of time and human effort to update and maintain, and the process is prone to inconsistencies due to human error.

These factors have a direct impact on a plantâ€™s safety. The American Chemistry Council stated that in 2023, failures related to SOPs inaccuracy ranked among the top two causes of process safety events

Chemical companies that maintain better SOPs can improve process safety. The maintenance process can be accelerated by adopting tools that use natural language processing and large language models, by:

- Using text analysis for more efficient SOPs review.
- Accept or reject recommendations for SOPs revisions.
- Auto-generate updated SOPs based on those accepted recommendations.

When human ingenuity and AI are combined, it leads to a 25% reduction in SOPs review time and a 60% improvement in review accuracy. Standardization of SOPs leads to clarity and consistency across an organization, improvement in plant safety compliance and a reduction in unscheduled downtime.

In this session you will:

- Understand how AI can contribute to improved process safety.
- See a demo of the Smart SOP Control Tower tool.
- Explore the benefits and use cases for Smart SOP Control Tower.

2024 GLOBAL CONFERENCE ON PROCESS SAFETY & BIG DATA

Integration of AI in PPS

CCPS EPSC

CSE CENTER OF SAFETY EXCELLENCE

Integration of AI in Process and Plant Safety - a paradigm change?

Jürgen Schmidt, CSE Center of Safety Excellence, Pfinztal

The World is Changing

Internet o Things (IoT)

Industry 4.0

Artificial Intelligence (AI)

Digitalization

Automation

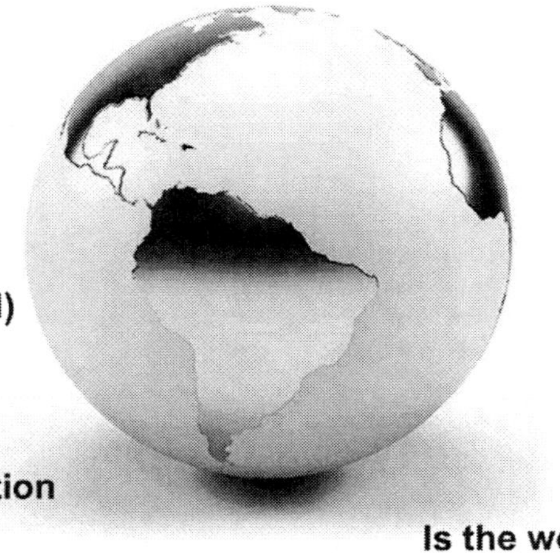

Is the world becoming safer?

2024 PROF. DR. JÜRGEN SCHMIDT | CSE CENTER OF SAFETY EXCELLENCE | PFINZTAL 8

AI and Big Data

- Biometrics
- Autonomus Driving
- Robotics
- Chatbots
- Large Language Processing

2024 PROF. DR. JÜRGEN SCHMIDT | CSE CENTER OF SAFETY EXCELLENCE | PFINZTAL 9

Protection of Processes and Plants

Technical countermeasures

- Safety devices
- Safety instrumented systems

- Proof of function
- Tested accuracy, reliability (type testing, SIL-level)
- Invariable over time
- Certified

Contrary to AI !

Decision Making

Math - Algorithm:
Procedure to find THE verifiable best solution

Engineering - Empirics:
Model to find a verified solution

AI - Heuristics:
Procedure to find ONE solution

Human - Heuristics:
Good solution
in case of uncertainty

Development of an AI model (heuristic)

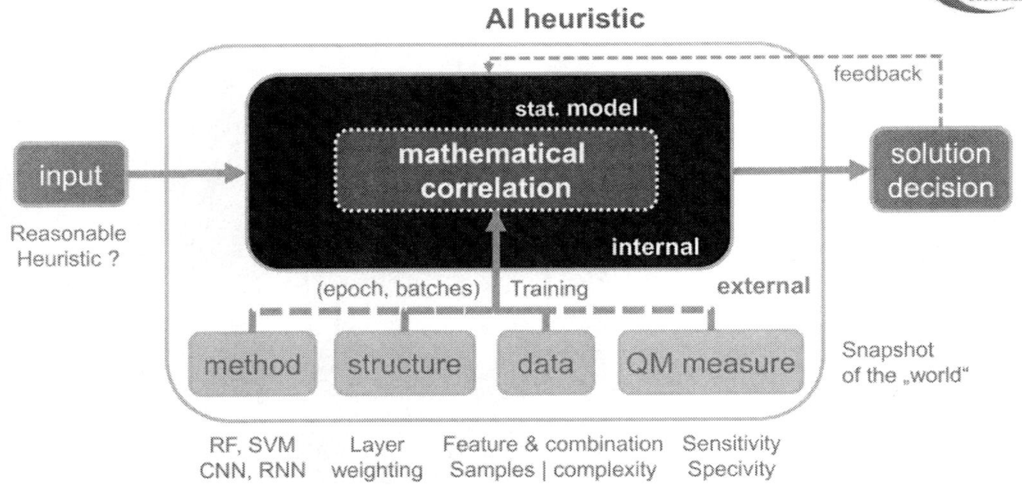

Decision Making

Single desicion
once in a time

Group based - statistic:
(credit worthiness)

Evaluation - conventions:
(grade, diagnosis)
accepted human process
explainable in human categories

AI Decision Making is NOT Comprehensible

CSE
CENTER OF SAFETY EXCELLENCE

AI Safety Vision

HAZOP **Operating license from authorities**

Process-Monitoring
Swarm-sensors

Smart p,T protection

Anomaly-detection

Root-cause
explanatory components

Test cycles:
risk- & opertation related

Multi-barrier redundancy & diversity

MetA-HAZOP CSE PhD project

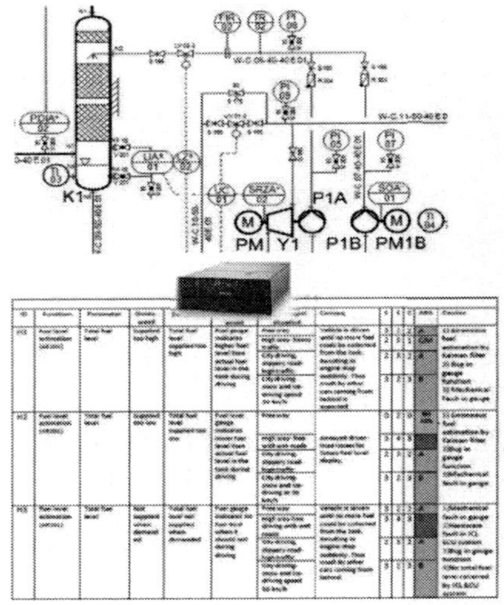

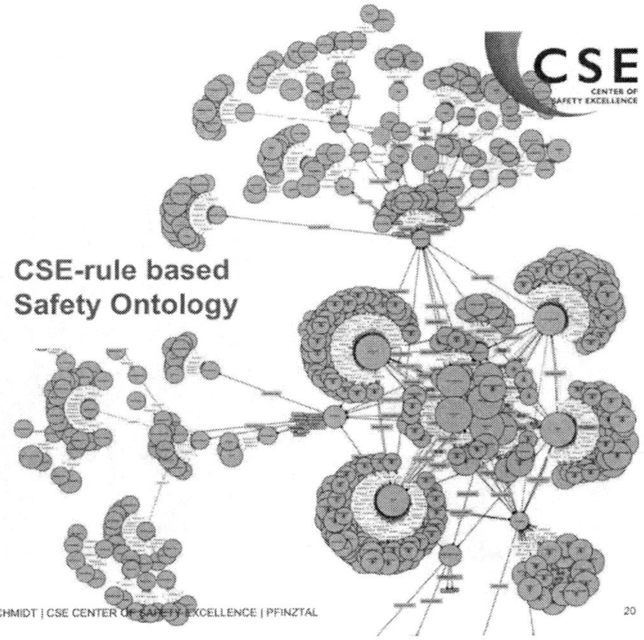

SISProof

High integrity diagnostics based on AI

AI-soft sensors
redundant, divers

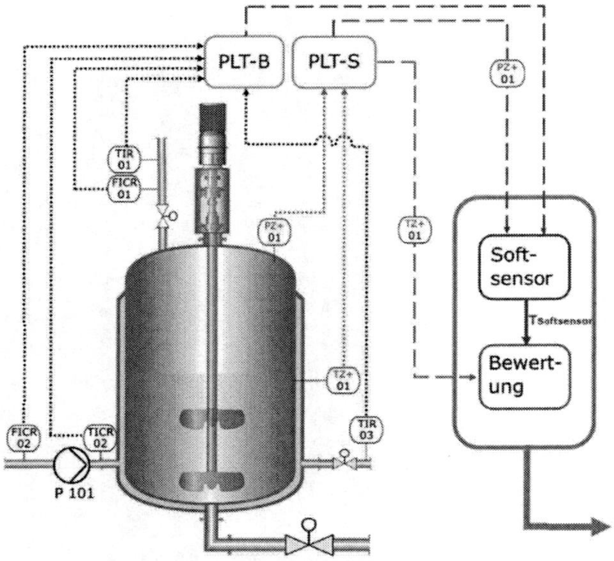

Hybrid Modeling

CSE
CENTER OF SAFETY EXCELLENCE

CSE PhD project

sRMC CSE PhD project

Anomaly Monitoring and Control

Multi-sensor
Deep Learning

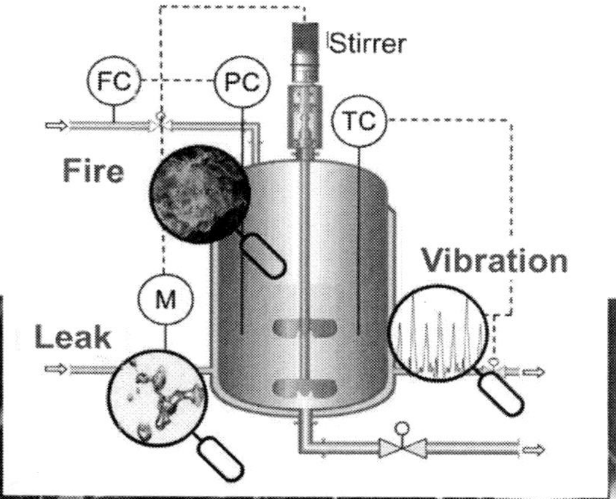

RiIM - AI CSE PhD project

Risk based
Integrity Management based on AI

Corrosion
Third Party Incidents
Natural Events

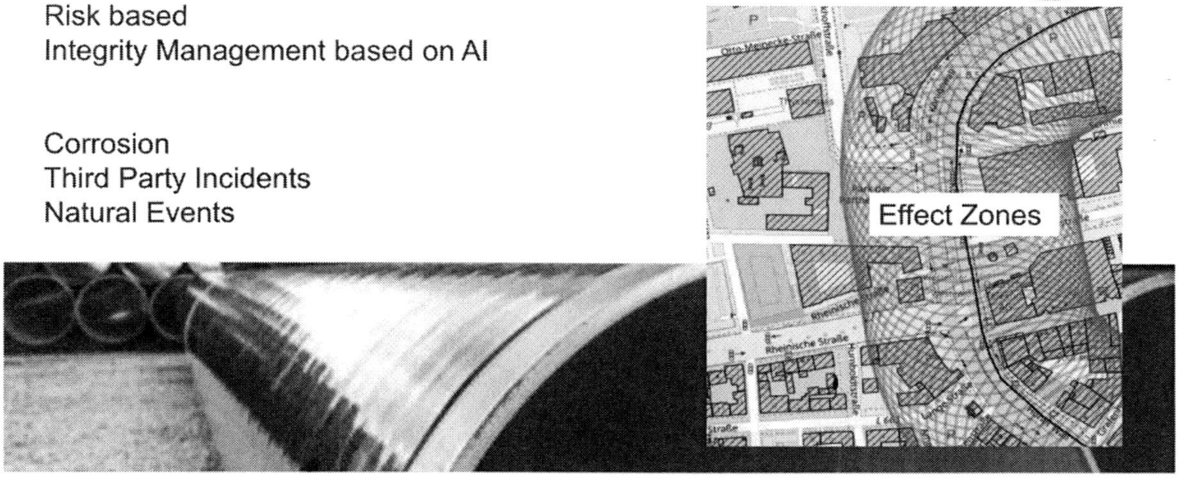

SmartHIP | SmOP - pressure protection

Zero Emission:
AI based Safety & Security

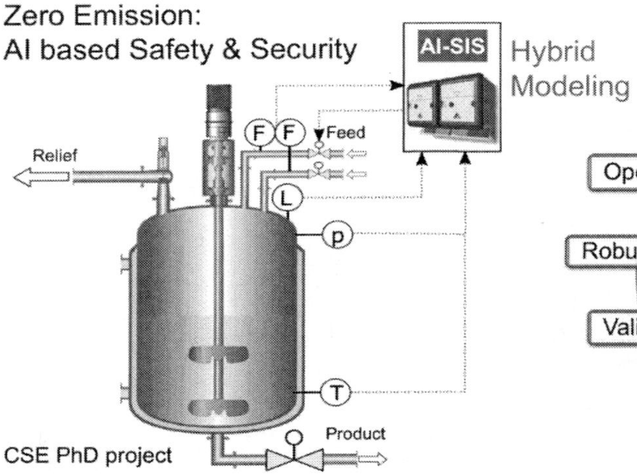

CSE PhD project

High Demand

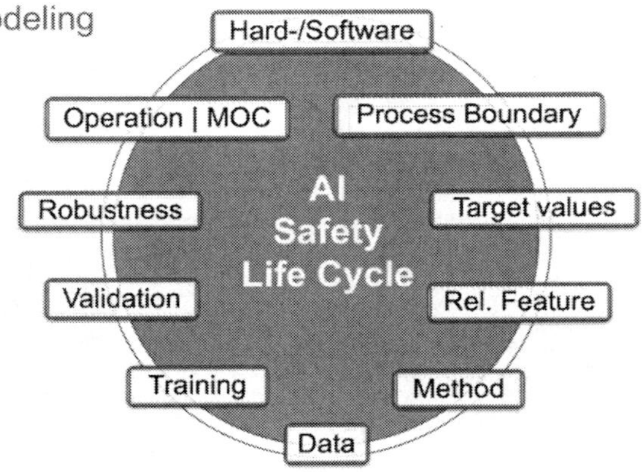

Can we rely on AI protection systems ?

What is the basis for trust to protect systems ?

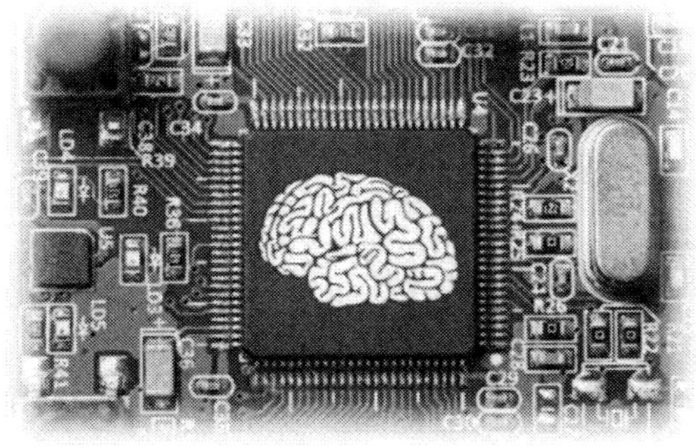

We Manage Human Errors

- Anxiety
- Lack / disorder of concentration
- Lack of responsiveness
- Insufficient situational awareness
- Decisions not made carefully
- Misunderstanding of instructions, contradictory instructions, ...
- Recklessness (ignoring warning signals)
- Lethargy
- Disregard for common practice
- Excessive demands or stress
- Lack of knowledge / skills
- Too strong belief in own competences
- Fatigue | Exhaustion
- Shock
- Injuries | Illness or pain
- Alcoholizm, drug intoxication or medication

Personnel Selection Process

Big 5 of Personality

- 5 core properties
- 3125 personality variations
- **Choise of personnel:** sufficient experience to assess skills & personality
  - » Culture, Age, Health
  - » Education (certificate)
  - » Training
  - » Personality

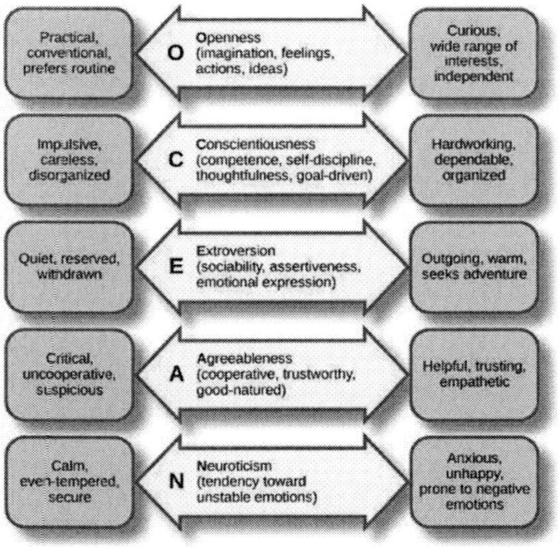

AI Selection Process - Challanges

- Accepted methods/procedures
- Completeness of features
- Data quality / availability
- New proof tests for individual models with stochastic nature
- Freezing of model status (learning on demand)
- Regulatory compliance
- User adoption
- ....

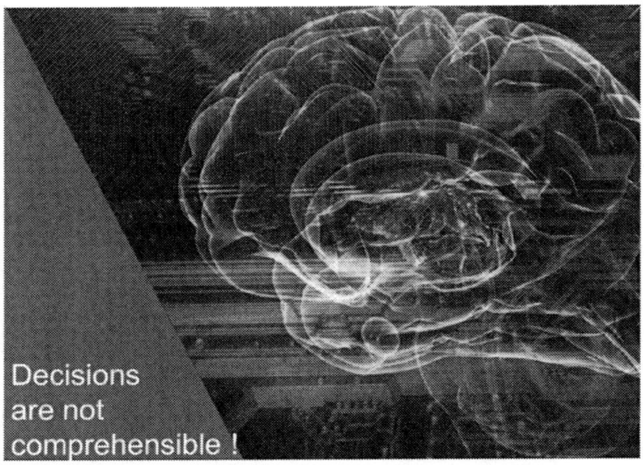

Integration of AI in PPS

AI safety-oriented life cycle:

- **Highest requirements**
- **AI SIL certification**
  - Safety proof
  - Transparency (trust)
  - Uncertainty assessment & robustness training
  - Verifiable implementation
  - Credibility of risk reduction
- Logic based system preferable
- Physics integration desirable

Integration of AI in PPS – a paradigm change !

AI in safety related applications:
HIGH RISK APPLICATION
according to AI act

- **Engineering | Documentation**
- **Assistance** system with human decision
- **Diagnostic** enhancement
- **Prevention** system to avoid pressure relief
- Secundary **mitigation** system
- **Training** system

2024 PROF. DR. JÜRGEN SCHMIDT | CSE CENTER OF SAFETY EXCELLENCE | PFINZTAL 30

Integration of AI in PPS – a paradigm change !

- Specific safety mindset required
- (Young) AI professionals are needed
  - R&D
  - Safety consultants
  - Authorities

Thanks for your attention

CSE
CENTER OF SAFETY EXCELLENCE

CSE Center of Safety Excellence, Pfinztal, Germany
Prof. Dr. Jürgen Schmidt
juergen.schmidt@cse-institut.de
0049 172 8763561

CSE
CENTER OF SAFETY EXCELLENCE

CSE
CENTER OF
SAFETY EXCELLENCE

Research and Education in Process and Plant Safety

CSE-INSTITUT.DE

HAZOP Assistant, the new co-pilot for process safety

Alfredo Haubold
Vysus Group
Avda. Corts Catalanes, 5-7
Sant Cugat del Vallès, Spain
alfredo.haubold@vysusgroup.com

Angel Casal
Vysus Group
Avda. Corts Catalanes, 5-7
Sant Cugat del Vallès, Spain
angel.casal@vysusgroup.com

Prepared for Presentation at
American Institute of Chemical Engineers
2024 Global Conference on Process Safety and Big Data
Frankfurt am Main, Germany
October 29-30, 2024

AIChE shall not be responsible for statements or opinions contained in papers or printed in its publications

HAZOP Assistant, the new co-pilot for process safety

Alfredo Haubold
Vysus Group
Avda. Corts Catalanes, 5-7, Sant Cugat del Vallès, Barcelona
alfredo.haubold@vysusgroup.com

Angel Casal
Vysus Group
Avda. Corts Catalanes, 5-7, Sant Cugat del Vallès, Barcelona
angel.casal@vysusgroup.com

Keywords: "Process Safety", "HAZOP", "Digitalization", "Digital Twin", "Digital HAZOP".

Abstract

The HAZOP Assistant software tool uses Multilevel Flow Modelling (MFM) language, a form of Explainable Artificial Intelligence (XAI), to improve the performance of Hazard and Operability (HAZOP) studies and other safety reviews. A functional (digital) twin of the facility or project is created based on the physical plant design and on operational experience built-in within the equipment libraries in the software. The tool accurately predicts how the process reacts based on failures or abnormal conditions, akin to deviations in a HAZOP. The tool was tested in a FEED project involving the design of a new gas compression platform. The paper presents the results of the case study, explains the insights drawn from the pilot project and the way forward for the software application.

1. Introduction

HAZOPs have been widely used in process safety for decades. They evaluate the safety and operability of a given process following a systematic identification of potential deviations from the intended operating conditions.

HAZOP has proven to be a very useful hazard analysis method to effectively eliminate or reduce process safety risks to a level that As Low As Reasonably Practicable (ALARP). Furthermore, HAZOPs are often the basis for other Process Hazard Analyses (PHA) such as Layer of Protection Analysis (LOPA) and Quantitative Risk Assessments (QRA) during design. During operations, there are other safety reviews such as Revalidation HAZOPs (re-HAZOP), Management of Change HAZOPs (MOC HAZOP), Root Cause Analysis (RCA), Pre-Start Safety Reviews (PSSR) and Operational Risk Assessments (ORA).

Therefore, the quality of the HAZOP is of utmost importance; missing hazards or making incorrect assessments of the risk and/or the required safeguards, could lead to process safety accidents.

HAZOPs still face significant challenges given that they rely heavily on human competence and experience during the team reviews. Human factors can reduce the effectiveness and accuracy of HAZOP, such as bias, fatigue, boredom, or groupthink. Casal (2022) [1] concludes in his paper that there is no guarantee of consistency, completeness and repeatability. In addition, HAZOPs are very time consuming and cannot be used directly as safety information is not structured i.e. causes for same scenarios are found in different parts of the HAZOP nodes.

As pointed out by Keddie, A. (2024) [2], several attempts are currently underway to develop software tools that incorporate Artificial Intelligence (AI) to improve HAZOPs, with four (4) solution providers at the time of writing this paper. The author is also aware of one other initiative between TNO and Shell but no information has been made publicly available regarding this development.

Vysus Group, in collaboration with Kairos Technologies, is one of these solution providers with the "HAZOP Assistant". The HAZOP Assistant tool has been used in four pilot studies in recent years to demonstrate value.

The latest one involved a digital HAZOP for a new gas compression platform during the Front-End Engineering Design (FEED) stage for a major Oil and Gas operating company in 2024.

The paper presents the results of the case study, explains the insights drawn from the pilot project and the way forward for the tool.

2. The HAZOP Assistant

2.1. Foreword

HAZOP Assistant is a cloud-based software which can be used to create digital twins of process plants. The digital twin is a functional model of the facility with qualitative mass and energy flows throughout the plant connecting all the process equipment as per the design information (P&IDs). The HAZOP Assistant includes equipment libraries for the process equipment which incorporate failure modes from publicly available sources as codes, standards, as well as Vysus experience performing HAZOPs and Failure Mode and Effects Analyses (FMEA).

With its qualitative reasoning capability, the HAZOP Assistant can generate prediction of possible causes and possible consequences for any given deviation from the design conditions as follows:

- If the model is queried with a deviation of a process parameter at any given location of the plant (e.g. more flow from a pump), the model identifies what the potential causes are which will result in the deviation. At the same time, the model will also predict what the effects are from this deviation, typically further deviations of process parameters both upstream and downstream. Some of these deviations can evolve into hazardous consequences and operability problems if not properly safeguarded;
- If the model is queried with a fault at any given location of the process (e.g. normally open valve accidentally closed), the model predicts what the effects are from the deviation, similar to the case above; and,
- If the model is queried with a consequence at any given location of the process (e.g. overpressure), the model predicts potential causes which will result in this consequence.

The language selected for the HAZOP Assistant is called Multilevel Flow Modelling (MFM). It is a qualitative language used to represent the behaviour of the process in a chemical or industrial facility (Lind, M. (2011) [3]) and was originally developed by the Technical University of Denmark (DTU).

MFM has been categorised as a type of AI known as Explainable AI (XAI). As opposed to other digital HAZOP tools which use generative AI, XAI does not learn by itself and only uses human knowledge previously loaded in the digital twin via the equipment libraries. This makes the reasoning of the tool completely transparent as it is deterministic, with a predictable behaviour which will always generate the same results for the same input values. This is not the case of other digital HAZOP tools which use generative AI.

As said earlier, libraries have been developed for all process equipment typically found in process facilities, onshore and offshore, as per as per ISO 14224 (2016) [4]. MFM enables the model to include all equipment in the facility to the required level of depth: it has capability to include sub-units, maintainable components and parts as per the previously mentioned standard.

The advantage of using qualitative functional modelling is that a single model can be used to represent different equipment with different objectives. For example, the functional model in the centrifugal pump equipment library can be used to represent any centrifugal pump regardless of size, duty, service and mission e.g. process pump, cooling fluid pump or firewater pump.

The functional relationships between equipment, and its components, and with the targets and hazards outweighs any database of words linking causes, consequences and safeguards such as the one created in a HAZOP with common HAZOP software. A key difference is that the functional twin can grow as more knowledge becomes available about the facility and can be modified and updated, as the facility evolves.

Another advantage of using a qualitative language is that the processing time can be significantly reduced, typically less than a minute for a medium-sized process plant.

2.2. *The digital HAZOP*

It is important to clarify that HAZOP Assistant has not been initially developed as a substitute of HAZOP workshops and the experience of its members; instead, the intent is for this software tool to help process safety experts in the thought process required to perform a HAZOP, so that no relevant events that would be expected to appear in a HAZOP record are missed. Furthermore, for future releases of the software, several features of standard PHA tools are being explored such as a node editor, admin modules for capturing session dates and attendees, reporting, etc. When implemented, the functionality of HAZOP Assistant could be such that standard PHA tools will not be required to meet all the requirements of HAZOPs.

The process starts by building a digital twin of the facility using the equipment libraries in the HAZOP Assistant. Afterwards, this equipment is connected as per the facility / project P&IDs. This is when all the control and safety components are added such as valves and sensors. Basic Process Control (BPCS) Loops and Safety Instrumented Functions (SIFs) are included in the model to enable the automatic identification of the relevant safeguards in case of process deviations. As an example, the figure below shows a simple digital twin of a process consisting of a degassing Column and Drum, a process pump, a condenser, a reboiler and a cooler.

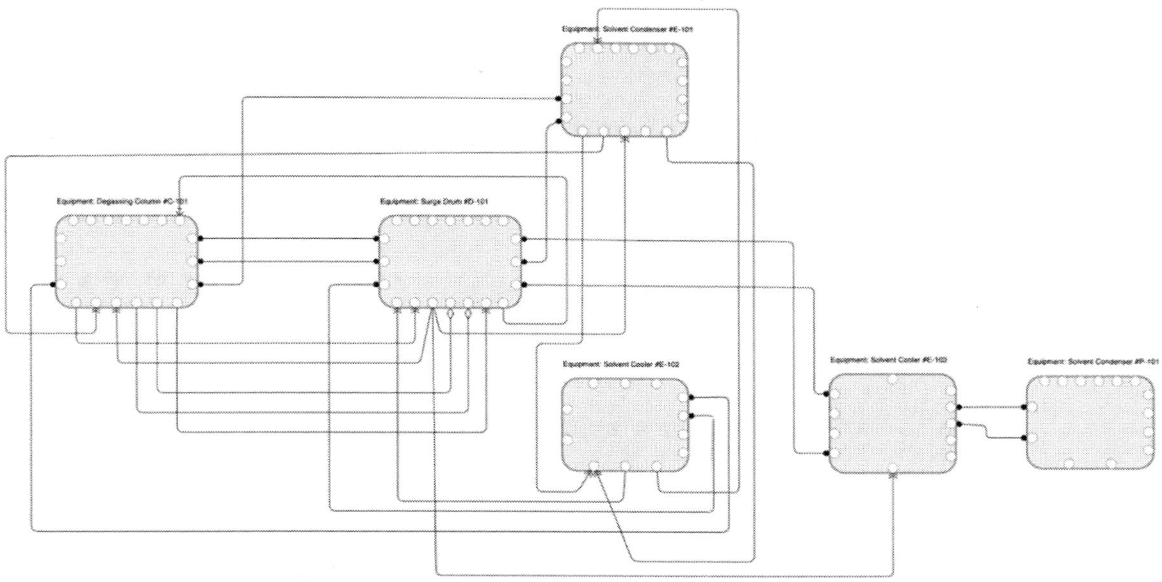

Fig. 1 — Functional twin assembly from sub-functions

After building the model, HAZOP Assistant uses it reasoning capability to generate a set of preliminary results which can be used to identify critical scenarios which must be addressed during the HAZOP workshop. A set of automatic rules are applied to ensure that every scenario is evaluated consistently in terms of likelihood, severity and risk. This ensures that consistency is applied throughout the study, regardless of its duration.

The next step is the HAZOP workshop. When HAZOP Assistant is used, thanks to the initial pre-workshop analysis the nodes, sessions, and required personnel can be better planned. etc. Most scenarios would have already been identified. The tool will easily address the bulk of readily predictable scenarios. e.g. valve mispositioning, external fire, etc., hence saving time and allowing the team to focus on critical and complex scenarios.

During the HAZOP workshop, the tool can also be used interactively as a co-pilot to ensure that no scenarios are missed as it´s reasoning capability is very fast.

In summary, HAZOP Assistant is a very powerful tool which allows users to create a digital twin of a facility, which can effectively improve the HAZOP process by guiding the HAZOP workshop, enhancing coverage of hazards with its equipment libraries, ensuring consistency and reducing time.

Furthermore, HAZOPs are often the basis for other Process Hazard Analyses (PHA) such as Layer of Protection Analysis (LOPA) and Quantitative Risk Assessments (QRA), hence these studies will also benefit from improved HAZOPs. During operations, there are other safety reviews such as Revalidation HAZOPs (re-HAZOP), Management of Change HAZOPs (MOC HAZOP), Root Cause Analysis (RCA), Pre-Start Safety Reviews (PSSR) and Operational Risk Assessments (ORA). As discussed by Casal (2023) [5], HAZOP Assistant can also efficiently support these processes during operation making use of the digital twin of the facility and providing further benefits. This is referred to as Process Safety Management (PSM) evergreening.

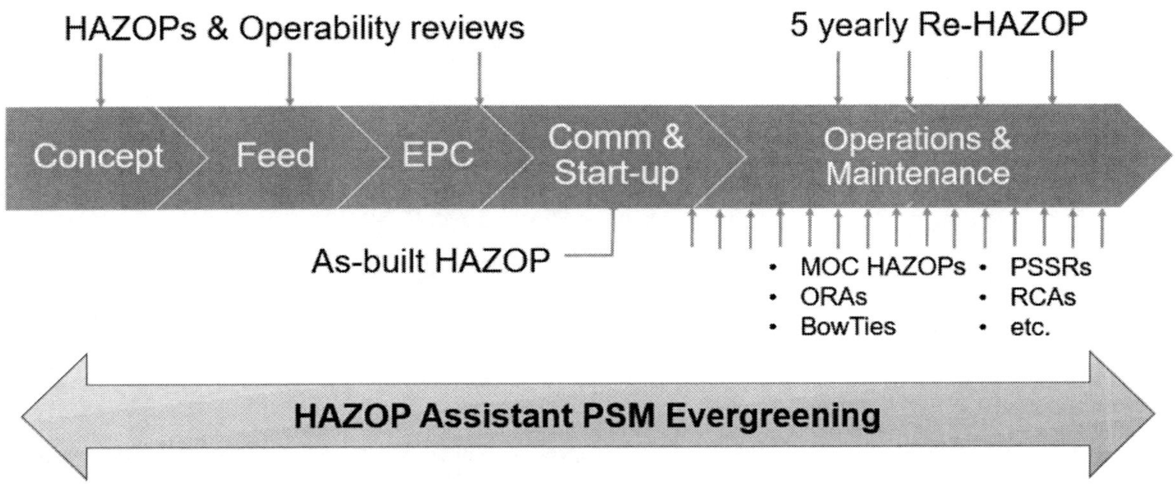

Fig. 2 — Typical PS reviews over the life of assets

3. Case study

3.1. The project

The HAZOP Assistant was applied in a pilot project early in 2024 involving the design of a new gas compression platform during the FEED phase.

The main objective was to undertake a digital HAZOP (dHAZOP) study for the Project to demonstrate the functionality of the HAZOP Assistant tool and its ability to improve the coverage of hazards and ensure repeatability and consistency of the process, whilst reducing the time and costs spent in HAZOPs.

A new Gas Compression Platform (GCP) is planned to be installed in an existing oil field as part of a program to enhance hydrocarbon recovery. The new GCP will receive the associated gas from nearby platforms and will gather, compress, dehydrate and export it gas via offshore pipeline to a facility onshore.

The new GCP has the following key process systems:

- Slug Catcher;
- First Stage Gas Compression (3 x 33.3%);
- TEG Contactor/Regeneration (2 x 50%);
- Hydrocarbon Dew Point Control (2 x 50%);
- Second Stage Gas Compression (3 x 33.3%);
- Fuel Gas system;
- Seawater/Tempered water system;
- Instrument Air Package;
- Nitrogen Package;
- Open/Closed Drains;
- Flare Headers; and,
- Utilities Distribution.

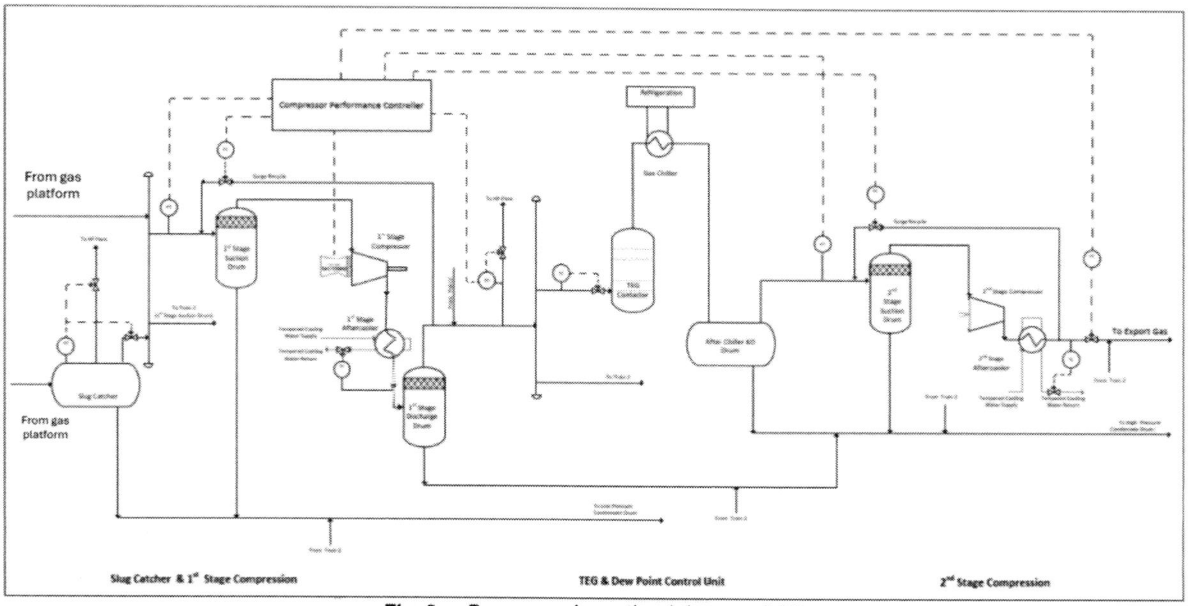

Fig. 3 — Process schematic of the new GCP

3.2. *Scope and limitations*

The main tasks in the scope of work were:

- Prepare a Terms of Reference (TOR) for the dHAZOP;
- Build a functional twin of the process based on facility information (P&IDs, PFDs and C&E);
- Desktop analysis using the HAZOP Assistant and preparation for the dHAZOP workshop;
- Facilitation of a dHAZOP workshop; and,
- Preparation of the dHAZOP Report.

The following systems were not modelled:

- Seawater/Tempered water system;
- Instrument Air Package;
- Nitrogen Package;
- Open/Closed Drains;
- Flare Headers; and,
- Utilities Distribution.

The dHAZOP considered:

- main process systems of the Project only. Utilities and vendor packages were excluded and treated as "black boxes";
- process safety hazards; the focus of the pilot was not operability issues;
- operations phase of the project only. Other phases such as fabrication, transport, installation, commissioning and decommissioning were excluded; and,
- normal operating mode only. Other operating modes including but not limited to start-up, shutdown, maintenance, etc. were excluded.

3.3. *Current limitations*

As the HAZOP Assistant uses generic libraries, project / site / reservoir specifics are therefore absent and need to be covered in the dHAZOP workshop. Examples include:

- Composition issues / changes;
- Impurities; and,
- Hydrates.

Similarly, plant wide specifics not directly connected with the equipment in the model are hard to cover in the functional twin. This would include the issues that are normally included in the so-called “Global Node” in a normal HAZOP. Examples include:

- Extreme weather events e.g. wind, waves, sandstorms, etc.;
- External impacts e.g. vehicles, ships, cranes, etc. However, external fires are included;
- Global power failure. Local power failure however is considered as all powered equipment is assumed to fail within the model;
- Ergonomics and human factors are excluded.

Another important feature of the model is that is qualitative, hence it cannot predict how high will be pressure, temperature, flow or level or how long will take for something to happen. However, most HAZOPs rely on simple qualitative evaluations performed by humans so this is not considered a real limitation, but an opportunity to enhance the HAZOP process significantly.

At this stage, the HAZOP Assistant can only identify automatically BPCS, Alarms, SIS and PSVs; other controls such as standard operating procedures, integrity management programs, sampling, etc. must be included manually in the model. Further, the HAZOP Assistant does not perform a detailed evaluation of the effectiveness of the safeguards at this stage e.g. evaluation of Independent Protection Layer (IPL) success criteria. However, it can visualize gaps in safeguarding allowing for easy identification of critical scenarios requiring additional protections.

3.4. Evaluation criteria

As mentioned earlier, the main challenges of HAZOPs are:

- **Consistency:** A HAZOP workshop normally lasts for several days, and its members make hundreds of assessments of scenarios every single day. It is difficult for the team to keep consistency with the evaluation of the risks throughout the entire HAZOP. Commonly, there will be different conclusions regarding causes, consequences, risk, required safeguards, etc. because people cannot remember exactly what was concluded earlier. Sometimes, weeks pass in between sessions.

- **Completeness:** Completeness is unarguably the main aspect that can affect the quality of a HAZOP. It relates to how many scenarios have been identified in the workshop in relation to how many there could be in the facility. The completeness of the study will depend on whether the right combination of competence and experience are present in the workshop, the adequate allocation of time for the reviews and an experienced chairman that leads the sessions. Even if the workshop team involves all the skills needed, the human cognition will still affect the results.

- **Repeatability:** The HAZOP study is done with a team workshop, based on humans going through certain questions. The results obtained from the study will differ greatly depending on the time of the day, the stress each individual has or the level of competence or experience. That makes the study unrepeatable. Even the same personnel and chairman cannot guarantee that if they repeat the study one week later, the results will be different.

- **Resources:** The study requires the presence of highly skilled personnel, with high knowledge of the facility and/or the design. HAZOP studies usually take several weeks to complete. This means that the study has a high time cost.

- **Usability:** PHA studies are connected. As said earlier, several of the PHA studies do use the HAZOP as its base. The connection between the different studies generates a need for the HAZOP to be as usable as possible. The more easily understood is the information in the study, the more efficient can be the related studies.

Therefore, these aspects will be used as criteria for evaluating the performance of the HAZOP Assistant during the pilot study.

It should be noted that the client performed a traditional HAZOP with a third-party consultant, however a copy of the study was not made available to Vysus. Hence, a full comparison of both approaches in respect to certain aspects was not possible.

4. Results

Prior to the dHAZOP workshop, relevant P&IDs were modelled in the HAZOP Assistant as per the agreed methodology.

4.1. Node definition and systems modelled

The HAZOP Assistant does not require defining nodes as all equipment is modelled holistically. The following table describes GCP systems and equipment included in the model.

Table 1 GCP systems and equipment

System	Equipment	Comments
Inlet facilities	Inlet facilities and slug catcher	
1st stage compression (3 x 33%)	1st Stage Compression 1st stage suction drum 1st stage compressor 1st stage aftercooler 1st discharge drum	Train 1 only
Gas dehydration (2 x 50%)	TEG Contactor Gas Chiller After Chiller KO Drum	Train 1 only. Propane refrigeration system included to show value even though design is preliminary
2nd stage compression (3 x 33%)	2nd stage suction drum 2nd stage compressor 2nd stage aftercooler	Train 1 only
Condensate collection system	Low Pressure condensate drum Low Pressure condensate drum pumps High pressure condensate drum High pressure condensate drum pumps	
Fuel gas system	Fuel Gas heater Fuel Gas KO Drum Fuel Gas filter	

For the generation of the model, P&IDs were used, along with two one-hour meetings with process engineers responsible for the design. These meetings were set to solve any doubts the modelers could have and make the digital twin as close to reality as possible.

4.2. Rules and preliminary analysis

Each scenario that has the potential to cause harm has been evaluated with respect of "Likelihood" and "Severity of the consequences" as per the automation rules. This evaluation is performed automatically by the HAZOP Assistant, reducing time, increasing consistency and avoiding errors. This is achieved by automating likelihood and severity selection in the risk matrix of the client based on a set of rules.

A sample of the main automation rules for likelihood and severity used are presented in the following tables. Since the risk matrix used during the study is confidential, the corresponding likelihood and severity category are not shown, but can be easily inferred.

Table 2 Sample of ruleset for Likelihood (Causes)

Cause	Type	Frequency (1/year)
BPCS instrument loop failure	Instrumentation	1.00E-01
Valve passing	Mechanical	1.00E-01
Regulator failure	Mechanical	1.00E-01
Heat exchanger tube rupture (i.e. internal leak)	Mechanical	1.00E-03
Pumps and other rotating equipment failure	Mechanical	1.00E-01
Pressure Safety Valve (PSV) opens spuriously	Mechanical	1.00E-02

Table 3 Sample of ruleset for Severity (Consequence) for LOC of flammable / toxic gas

Applies to	Deviation	Effect	Consequences (flammable & toxic gas)
Static equipment e.g. vessels, drums, separators	High pressure	Overpressure, potential LOC	Toxic exposure, Jet fire if ignited, multiple fatalities, equipment damage.
Compressors (deadheading case)	High pressure in Compressor Discharge	Compressor deadheading, overpressure, potential LOC. Compressor Surge.	Toxic exposure, Jet fire if ignited, multiple fatalities, equipment damage.
Static equipment e.g. vessels, drums, separators	Low Pressure	Low operating pressure, operability issue only.	Operability issue only.
Compressors (starve case)	Low Pressure in compressor suction	Compressor surge, equipment damage, and potential LOC.	Toxic exposure, Jet fire if ignited, multiple fatalities, equipment damage.
Equipment with potential overheating	Specific high temperature	Overtemperature, potential LOC	Toxic exposure, Jet fire if ignited, multiple fatalities, equipment damage.
Rest of the cases	High temperature	High operating temperature, operability issue only.	Operability issue only.

It should be noted that in the sample of rules in the above table, consequences are only given for flammable and toxic gas service. Rules were developed for all possible services e.g. gas, liquid, toxic, flammable, etc.

4.3. Pre-workshop results

There were 7,266 scenarios in the model; that is 7,266 unique combinations of any given cause in any given equipment, and consequences in that equipment or in any other equipment. The overview of total causes and scenarios is presented in the following table.

Table 4 Overview of causes and scenarios (total)

Equipment	Causes	Scenarios	% Scenarios
1st Stage After Cooler Train 1	20	519	7%
1st Stage Compressor Train 1	12	669	9%
1st Stage Discharge Drum Train 1	14	318	4%
1st Stage Suction Drum Train1	13	373	5%
2nd Stage After Cooler Train 1	23	191	3%
2nd Stage Compressor Train1	12	128	2%
2nd Stage Suction Drum Train 1	13	172	2%
Accumulator	6	17	0%
After chiller KO Drum	8	293	4%
Condenser	14	102	1%
Fuel Gas Filters	20	281	4%
Fuel Gas Heater	11	275	4%
Fuel Gas KO Drum	22	751	10%
Gas Chiller	16	432	6%
Slug Catcher	19	858	12%
High Pressure Condensate Drum	27	475	7%
High Pressure Condensate Drum pumps	8	53	1%
Low Pressure Condensate Drum	16	165	2%
Low Pressure Condensate drum pumps	9	60	1%
Refrigerant Suction Drum	10	38	1%
Refrigeration Compressor Train 1	11	130	2%
Refrigeration Economizer	7	89	1%
TEG Contactor	23	877	12%
TOTAL	334	7,266	100%

The large number of scenarios generated by the model is caused by the propagation of deviations to the operating process variables (e.g. high and low pressures, temperatures, flows and levels, etc.) from equipment to equipment. These deviations are generated due to local causes in any given equipment; in addition to potentially creating local scenarios, these deviations also propagate to other equipment, potentially creating additional scenarios there. Not all these scenarios will necessarily result in safety or environmental consequences. In fact, in most cases, these will only be operability issues due to process conditions being "abnormal", but not exceeding the safe operating envelopes of equipment.

Furthermore, the model also identifies deviations which are propagated to the boundary limits of the model: this is where the equipment within the model is connected to another system which is not in the model. The number of these scenarios is very large as there are several interfaces between the equipment within the model and other systems such as pipelines to/from other platforms, cooling water, flare header, TEG regeneration system, oil shipping pumps, fuel gas supply to consumers, gas export pipeline, etc.

Table 5 Overview of causes and scenarios (in-model)

Equipment	In-Model Scenarios	% Scenarios
1st Stage After Cooler Train 1	201	8%
1st Stage Compressor Train 1	328	13%

Equipment	In-Model Scenarios	% Scenarios
1st Stage Discharge Drum Train 1	114	5%
1st Stage Suction Drum Train1	193	8%
2nd Stage After Cooler Train 1	98	4%
2nd Stage Compressor Train1	207	8%
2nd Stage Suction Drum Train 1	54	2%
Accumulator	8	0%
After chiller KO Drum	88	4%
Condenser	36	1%
Fuel Gas Filters	11	0%
Fuel Gas Heater	28	1%
Fuel Gas KO Drum	29	1%
Gas Chiller	207	8%
Slug Catcher	297	12%
High Pressure Condensate Drum	20	1%
High Pressure Condensate Drum pumps	22	1%
Low Pressure Condensate Drum	14	1%
Low Pressure Condensate drum pumps	20	1%
Refrigerant Suction Drum	29	1%
Refrigeration Compressor Train 1	90	4%
Refrigeration Economizer	46	2%
TEG Contactor	316	13%
TOTAL	2,456	100%

In summary:

- There were 7,266 scenarios in the model; that is 7,266 unique combinations of any given cause in any given equipment, and consequences in that equipment or in any other equipment;
- Of these, 2,456 scenarios identified within the GCP model; and,
- The rest (4,810 scenarios) propagate at the interfaces of the model from causes originating in the model. These can be understood as scenarios which propagate outside the boundary limits of the model.

4.4. Workshop venue, date and attendees

The objective of the dHAZOP workshop was to evaluate the results generated by the tool and record any scenarios that were not considered. This evaluation served afterwards the purpose of improving the model created, as all the discrepancies were added onto the model.

The dHAZOP workshop was held in USA from 15th to 18th January 2024.

The first session began with a safety briefing, followed by a round-the-room self-introduction by each of the dHAZOP participants. A brief 'Safety Moment' was also presented at the start of the workshop. The workshop was attended by a small multidisciplinary team with appropriate knowledge and experience both of this project and similar projects.

4.5. Workshop proceedings

The dHAZOP workshop was performed as per the agreed methodology in the TOR.

The key steps are summarised below:

- The dHAZOP facilitator identified a node/system for consideration;
- The lead process engineer briefly described the process and operations related to the node under consideration;
- The dHAZOP facilitator selected a consequence. The consequence as well as the causes which were credible to lead to the scenario selected were evaluated by the team using the HAZOP Assistant analytical and visualization tools (HAZOP Analyzer, PID view, Tree view). If some scenarios required a specific comment, the comment was added in the model;
- The dHAZOP team validated the risk ranked (likelihood and consequences) for all scenarios;
- All identified safeguards were evaluated for each scenario using the HAZOP Analyzer. If additional controls / safeguards were required, an action was raised;
- When the consequence was complete, the next consequence was selected, and the process was repeated until all consequences were reviewed;
- Then the facilitator asked if there were additional scenarios in that equipment for completeness. If required, these were added to the model outside the workshop. When the review of the equipment was competed, the next equipment was selected, and the process was repeated until all equipment were reviewed.

5. Results and analysis

The dHAZOP process was assessed by Vysus Group against the evaluation criteria previously stated.

<u>Consistency</u>

As per the dHAZOP Report, the HAZOP Assistant identified over 7,000 scenarios. These scenarios were classified by the tool and had an automatic risk ranking depending on the service (gas, condensate, glycol, etc.). There were several rules included in the tool to automatically apply a likelihood on the causes or a severity on the consequences.

Including these rules into the model allowed the results to be automatically categorized into high, medium or low risk scenarios. Hence, the evaluation of the scenarios from the point of view of likelihood and consequences was consistent throughout the study.

<u>Completeness</u>

Regarding the completeness of the study, the digital twin was pre-loaded with the process libraries. These libraries contain publicly available knowledge on failure modes, such as the ISO 14224 and the API 14C. During the dHAZOP workshop it was found that all the critical scenarios were included. Some operability scenarios were missing, which were afterwards fed into the digital twin. The completeness of the models is a quality that will be improved with further use of the tool, as any scenarios not found by the tool can be added afterwards to the digital twin, making them available further on.

The deterministic nature of the tool also ensures that every output can be explained, which is a significant advantage over other tools which use generative AI i.e. black box.

Repeatability

After the digital twin for the facility was made, the reasoning of the dHAZOP is unchanged: the results will be always the same since they are embedded in the model. This also means that if a HAZOP is to be done in a similar facility, the same model can be used as a base. This will lead to great reduction in the resources spent for further studies.

Resources

As said earlier, a copy of the traditional HAZOP for the facility was not made available to Vysus, hence the exact time and systems covered are unknown. Thus, it is not possible to assess the performance of the dHAZOP compared with the traditional HAZOP. Based on experience, a traditional HAZOP for the new GCP would be in the range of 10-12 days. The dHAZOP workshop was performed in 4 days.

Even though it was possible to perform the dHAZOP in 4 days, the modelling time was approximately 1 month for two (2) modellers. The value on resource reduction would come with further studies such as re-HAZOPs, MOC HAZOPs, etc, since the model is already built. The functional digital twin can be re-used at any moment, reducing the time spent in those studies.

Usability

The usability of the tool was assessed from two perspectives. The usability of the tool during a dHAZOP focussing on User Interface (UI), and the usability of the study and the digital twin afterwards.

From the perspective of the usability of the tool during a dHAZOP, the UI of the dHAZOP has room for improvement. Since nodes and deviations were not implemented on the tool at the time of the project, navigating data proved to be overwhelming at times for the workshop team, as there was too much information to digest. Nodes and deviations help humans divide the complexity of the plant into smaller pieces which are better understood and brainstormed upon. From this project, it was realized that the user interface of the product needed improvements to optimize its usage as a HAZOP tool.

Since the digital twin is built and the information therein is presented in a coherent and structured manner when queried, the information can be readily used at any moment, unlike traditional HAZOPs which are rarely consulted due to the information presented there not being systematically organised. This will allow other processes to re-use the digital twin such as safety trainings, re-HAZOPs, MOC HAZOPs, FMECAs, RCAs, PSSRs, etc. This concept is called PSM Evergreening and has as an objective the complete availability of all the Process Safety Information (PSI) at any time, making the studies "live".

In summary, it can be concluded that 3 of the 5 main aspects evaluated as part of the case study were positive: HAZOP Assistant has improved the usability, the repeatability and the consistency of the study. The resource reduction and the completeness aspects could not be fully evaluated.

6. Future work

The lessons learned from the pilot study led to the identification of several activities that are being researched currently to improve HAZOP Assistant. The main ones are summarised below:

- Model automation: The creation of the models was the most time-consuming project activity. One of the ideas raised was to create the digital twin directly from smart P&IDs. Once the whole process is automated and tested, the model will be able to do faster HAZOPs, reducing significantly the time needed from modellers.
- Scenario generation: During the study it became clear that the models were overly complicated. The creation of models aimed to replicate the process as accurately as possible, these models led to the identification of over 7,000 scenarios, most of them being duplicates (parallel scenarios). As a result, the modelling guidelines will be revised to simplify the models which will result in less "noise".
- UI improvements: The UI of the tool at the time of the HAZOP workshop was suboptimal. For each scenario, the UI shows causes, consequences, safeguards, plus a field that is available for the introduction of text associated to each scenario. However, the information in the UI was presented in a different order than how it is shown in a HAZOP record. At the dHAZOP, it was obvious that filters for deviations and nodes were needed to better guide the HAZOP team. Consequently, the inclusion of these filters to the UI was identified as a key improvement to the software tool, along with other UI improvements.
- Consequence evaluation: The need to automate the identification of breaches and increase the accuracy of the model to narrow the list of safety significant scenarios is considered a key development. However, this would require assigning quantitative safe design limits to the equipment (i.e. design pressures, temperatures, etc.) which would push the limits of the tool as one of its great advantages is its qualitative nature.

7. Conclusions

Several attempts are currently underway to develop software tools that incorporate Artificial Intelligence (AI) to improve HAZOPs.

Vysus Group has developed the HAZOP Assistant which uses Explainable Artificial Intelligence (XAI) algorithms to create digital twins of process plants and which can predict causes and consequences from deviations from process parameters.

The tool was tested in a FEED project involving the design of a new gas compression platform in early 2024. It can be concluded that of the five (5) main aspects evaluated as part of the case study, three (3) were positive (usability, repeatability and consistency). The resource reduction and the completeness aspects could not be fully evaluated as a copy of the traditional HAZOP study was not made available to Vysus.

The pilot study generated valuable feedback which is being implemented in the tool to ensure that the tool can support all aspects needed in order to improve HAZOP studies by overcoming its challenges.

8. References

[1] Casal, A., & Dr Seccombe, P. W. (2022). Is the time right to digitalise the HAZOP? Hazards 32.

[2] Keddie, A. (2024). Accelerating Adoption of PHA AI Tools, Techniques to assist suppliers and users to rapidly assess and adopt existing Digital Tools, HAZ34_006_KEDDIE, Hazards 34.

[3] Lind, M. (2011). An introduction to multilevel flow modeling. Nuclear Safety and Simulation, 22-32.

[4] ISO 14224:2016 Petroleum, petrochemical and natural gas industries — Collection and exchange of reliability and maintenance data for equipment, 3rd Edition.

[5] Casal, A. (2023). PSM Evergreening with Artificial Intelligence. SPE-216445-MS, ADIPEC.

Practical Application and Validation of Enhanced Acoustic Leak Detection through Data Augmentation

Conference

Global Conference on Process Safety and Big Data

Year

2024

Proceeding

2024 Global Conference on Process Safety and Big Data

Group

General Program

Session

Applying AI / Case Studies

Time

Tuesday, October 29, 2024 - 12:35pm to 1:00pm

Authors

Quick, D. - Presenter
Leak detection in industrial plants is critical for safety and environmental protection. Traditional methods have limitations, especially in noisy environments. Background noise is a significant challenge in industrial leak detection. Training with real industrial noises on-site is impractical but crucial for effective detection. Background noise affects the Signal-to-Noise Ratio (SNR), which is critical for successfully implementing an acoustic leak detection system. Lower SNR means background noise can easily mask leak signals, complicating detection. Strategies to address this include transforming acoustic signals using Fourier Transformation and analyzing patterns through autocovariance and autocorrelation. Additionally, CNNs trained on leak sounds without background noise perform poorly in noisy environments, whereas those trained with background noises show improved accuracy. Real-world industrial environments, where background noises are louder and more varied, require system adaptations to each specific setting.

A previous not published paper (Quick et al., (2024)) developed an enhanced acoustic leak detection system using data augmentation to overcome background noise challenges, employing convolutional neural networks (CNN) and a mixup technique. The developed system uses two datasets: one for leak noises recorded in a laboratory and another for background noises recorded on-site. These datasets are combined using a mixup technique, creating realistic training data that mirrors the acoustic environment of industrial settings. The method leverages Short-Time Fourier Transformation (STFT) spectrograms and trains a CNN to distinguish between leak and non-leak sounds. The proposed method was tested using the IDMT-ISA-Compressed-Air (IICA) dataset, demonstrating detection accuracies of 92.4% for ventleaks and 86.4% for tubeleaks, outperforming other methods. The depth of the network architecture significantly impacted detection accuracy, with deeper networks like ResNet18 providing better results. The method showed improved performance over existing approaches, particularly in noisy environments. While laboratory tests have shown promising results, real industrial environments present unique challenges that cannot be fully replicated in controlled settings. Background noises in actual industrial plants are more varied and complex, which can affect the performance of the detection system. Therefore, it is essential to test and validate the enhanced leak detection method in real-world industrial plants to ensure its reliability and robustness in diverse operational conditions.

To validate the proposed methodology, a series of tests will be conducted at industrial plants. A specialized test facility has been constructed to generate industrial leak noises. This facility includes a pressure reservoir capable of reaching up to 7 bar and is equipped with two G.R.A.S. 146AE 1/2" microphones positioned up to 8 meters from the leak source. Additionally, the facility features a Coriolis mass flow measurement tool and interchangeable nozzles to simulate different leak geometries. These nozzles are 3D printed using stereolithography to ensure minimal tolerance.Background noises were recorded over two measurement days at Industrial Plant X and Industrial Plant Y. Furthermore, leaks were artificially generated and recorded at these plants. The same leak geometries were also reproduced under laboratory conditions for comparison.

The gathered laboratory leak data will be mixed with the background noises recorded from the industrial plants. These augmented superpositional data will undergo bandpass filtering and transformation into spectrograms via Short-Time Fourier Transformation (STFT). The input size of the spectrograms will be (384, 384) for the convolutional neural network used for training and validation. Subsequently, the recorded leak data from the industrial plants will be utilized to test the network's accuracy, recall, and precision, verifying the methodology. The findings from these tests will be presented at this conference.

The Energy Sector Is Using State-of-the-Art Methane Leak Detection Technology to Reduce Emissions

Conference

Global Conference on Process Safety and Big Data

Year

2024

Proceeding

2024 Global Conference on Process Safety and Big Data

Group

General Program

Session

Applying AI / Case Studies

Time

Tuesday, October 29, 2024 - 1:00pm to 1:25pm

Authors

Bunegar, S. - Presenter, GrandPerspective GmbH
Braun, R., Grandperspective Gmbh
Methane emissions are a significant contributor to climate warming, with the European Commission (EC) stating that they account for around a third of all current climate warming. Research revealed that over a twenty-year period, methane has a global warming potential many times greater than carbon dioxide. All over the world pioneering petrochemical plants are already taking decisive steps to limit methane emissions. However, conventional sensors are often not capable of accurately detecting, quantifying and monitoring methane emissions.

Fourier transform infrared (FTIR) spectroscopy, exemplified by the scanfeldÂ® system, which is the brainchild of Grandperspective GmbH, a Berlin-based company, directly tackles the limitations of conventional sensors. Unlike traditional monitoring systems, it is able to accurately detect, identify, and quantify a wide range of gases, including methane. Our presentation will explain in detail how FTIR spectroscopy technology provides more accuracy and visibility than traditional systems.

In February 2024, a series of controlled-released experiments, which were validated by a renowned independent research institute, proved that Grandperspectiveâ€™s remote sensor technology was able to detect methane emissions at leak rates of only 100 grams per hour over a distance of at least 250 metres in real-life conditions. In total, Grandperspectiveâ€™s team conducted over 80 assessments experimenting with different flow rates and wind speeds, across a range of distances.

The results of these tests are a significant moment for the industry, as by scientifically proving that it is possible to detect and quantify emissions at extremely low detection thresholds. Using FTIR remote sensing technology for the first time, the sector has a set of tools that can help it to considerably reduce emissions.

The limitations of traditional methane detection monitoring systems highlight the need for advanced technologies like FTIR spectroscopy as these not only provide accurate measurements but also offer solutions to a myriad of different methane emission â€“ and other gas â€“ detection scenarios. As regulations tighten, this pioneering technology could prove an indispensable tool in significantly detecting, tracking, monitoring, and ultimately, reducing methane and other gas emissions.

Uses of AI for HAZOP / PHA

Conference

Global Conference on Process Safety and Big Data

Year

2024

Proceeding

2024 Global Conference on Process Safety and Big Data

Group

General Program

Session

Applying AI / Case Studies

Time

Tuesday, October 29, 2024 - 12:10pm to 12:35pm

Authors

Baybutt, P. - Presenter, Primatech Inc.
Artificial intelligence (AI) provides a variety of functions, such as prediction, classification and natural language processing, that can be used in the performance of hazard and operability studies (HAZOP) and process hazard analysis (PHA). This paper describes how AI can be used to support PHA studies. Pros and Cons are addressed and practical examples are provided including the use of AI classifiers to support the generation of HAZOP / PHA metrics and analytics using big data, and the use of generators to automate the completion of HAZOP / PHA worksheets. Lessons learned from the implementation of AI for HAZOP / PHA are described.

Conceptualizing Pressure Relief Requirements in the Design Phase: A Modeling Approach Based on Historical Data

Conference

Global Conference on Process Safety and Big Data

Year

2024

Proceeding

2024 Global Conference on Process Safety and Big Data

Group

General Program

Session

Applying AI / Case Studies / Practical Examples

Time

Tuesday, October 29, 2024 - 3:45pm to 4:00pm

Authors

Faulk, N. - Presenter, Siemens Energy AG
Datta, A., Siemens

In the petroleum refining industry, the price of crude oil and its refined products are highly volatile; influenced by global, regional, and local supply, demand, and regulatory changes. To be economically viable, the refinery must have the ability to transition efficiently and safely among various crude blends and throughput rates to their various units to optimize operation, economics, and consumer market demands. One critical requirement is evaluation of the impact of these potential feed and throughput changes, to determine the adequacy and impact on the existing relief and flare systems. Current (time-consuming) practices include involving pressure relief systems engineers to revalidate the proposed changes on the existing pressure relief and effluent handling systems.

This paper discusses the development of a model that uses data from hundreds of refinery units' pressure relief studies. The model provides a set of correlations between most processes (operating conditions, feed blends, product properties) and their overpressure relief requirements, based on industry standards and best practices. These correlations provide a powerful tool to determine key operating and safety parameters that control the pressure relief system sizing basis, *without the need for detailed pressure relief calculations*, during the early conceptual stage of any OPEX or CAPEX project.

Maximising the Value of Existing Process Safety Data: Three Case Studies from Empirisys

Conference

Global Conference on Process Safety and Big Data

Year

2024

Proceeding

2024 Global Conference on Process Safety and Big Data

Group

General Program

Session

Applying AI / Case Studies / Practical Examples

Time

Tuesday, October 29, 2024 - 2:45pm to 3:15pm

Authors

Carroll, G. - Presenter
Walters, M.
Vast amounts of data are being under-utilised in high-hazard industries. At Empirisys, we combine our process safety and data science expertise to maximise the value of existing data, ultimately to support - and not replace - decision-making to manage risk. The following three case studies exemplify our innovative approach: *Detect* connects operational and organisational data sources to measure and track Performance Influencing Factors (PIFs); *Boost* uses Large-Language Models (LLMs) to streamline and increase the utility of safety observations; and our enhanced barrier model offers new understanding of operational risk through data modelling and visualization.

Detect: Tools to Avoid Human Error

Detect is an application which connects to operational and organisational data sources, including control of work, maintenance, and incident applications. Detect is being developed in collaboration between Empirisys and BP and has been extensively reviewed and tested by industry experts and academic research. Detect uses a machine learning algorithm to identify underlying causes contributing to incidents, commonly known as Performance Influencing Factors (PIFs).

PIFs are conditions that influence an individual or team's performance in safety-critical environments. PIFs can encompass a wide range of elements, including safety culture, leadership, equipment reliability, training, procedures, communication, and many more. The goal of Detect is to help users to understand and manage these factors that are essential to ensuring safe and efficient operations in high-risk industries. Each PIF is calculated by averaging the score of their respective sub-components, which we call Markers. Each PIF score is built up of these individual Markers. The relevance and weight of each Marker to a specific PIF is calculated through a machine learning algorithm, which associates the impact of Markers on the likelihood of incidents by analysing past incidents.

The strength of Detect comes from the fact that PIFs and Markers can easily be monitored in the same tool. The machine learning algorithm helps to determine which Markers are of interest (i.e., there is a relationship between them and incidents at the same asset), and which Markers have relatively little impact. This ensures that only relevant data is monitored, and Markers that are useful indicators of incidents are brought into focus. It is important to note that a causal relationship between the markers and incidents is not assumed. The algorithm developed allows for greater understanding of the relationship between each marker and incident occurrence. This relationship is rarely linear; therefore, a RAG (Red, Amber, Green) status is utilised to indicate at what values of a marker incidents are more likely to happen. This facilitates monitoring of each Marker in a bespoke, statistically validated manner, rather than using a one-size fits all approach.

Boost: Utilising NLP & AI to extract insight from observation datasets

Boost is an AI enhanced Observation collection, analysis and reporting tool developed by Empirisys to enhance the value of observations. Boost ingests observation data and provides a prediction for relevant supplementary information using a combination of machine learning and large-language models. The analytical modelling pipeline ensures that consistent, complete, high-quality observation data can be utilised as a process safety indicator.

Core functionality of Boost was presented at the Global Conference on Process Safety and Big Data in 2023. This presentation will explore a selection of analytical techniques to maximise data value and utilization that have been developed in the elapsed months. Boost development has taken a data-centric approach to improve data quality of available observation data. This, combined with considered selection of appropriate machine learning techniques, has improved the quality of analysis insight from the existing data.

Data-centric model development focuses on increasing the quality of the input to a model to improve performance, rather than the â€˜traditionalâ€™ machine learning approach of attempting to devise a model that can â€˜replicateâ€™ a training dataset. This approach has been applied to a supervised machine learning classification task and has increased classification accuracy significantly. To translate this approach to the broader classification tasks that utilise Large Language Models (LLMs), Empirisys are supervising a Masterâ€™s project topic focused on identifying measures of intrinsic observation quality and reliability. For a subjective, text-based classification task, understanding the intrinsic quality of the input data will guide selection of targeted analytical techniques and elucidate valuable information that can be used to encourage high-quality observation submission and accurate analysis. This project utilises a dataset of observations labelled by subject matter experts, in a close collaboration between process safety experts and data scientists. Pertinent findings will be included in the presentation.

In addition, an alternative view of the observation data has been developed: unsupervised topic modelling to identify clusters of similar text with no prior injection of industry knowledge. This technique uses LLM text embeddings to identify patterns within observation datasets as a retrospective analysis. This is used to find similar observations, validate other classification tasks and aggregate information into meaningful sectors. Surfacing insight from observation data in this manner highlights the value that can be generated from existing data sources.

Barrier Model: The Power of Data Modelling & Visualization

Evidently, there are beneficial use cases for AI to unlock hidden insight. However, AI is one of many tools at the disposal of the data scientist and used only when it adds value. The enhanced hardware barrier model, developed in collaboration with Anasuria Operating Company (AOC), shows how restructuring data and visualising this in a new, intuitive way can produce great results.

The objective was to create useful insight from existing process safety data to support decision-making to manage risk. The work was split into two phases. In Phase 1, performance standards (SECEs) were mapped to Major Accident Hazards (MAHs) using the AOC safety case. In Phase 2, equipment was mapped to SECEs and MAHs. After a data quality assessment, risk criteria were established for Process Safety Performance Indicators (PSPIs) before visualising the barrier model insights in Power BI.

Management of process safety requires understanding of the status of people, process and plant. The focus of this work was on process and plant. Existing PSPIs were used to measure the integrity of hardware barriers and SECEs, as well as MAHs and their related Quantitative Risk Assessment (QRA) Events. Vulnerabilities to people were understood via AOCâ€™s competence management processes.

Prior to engagement with Empirisys, AOC monitored PSPIs, however only at the barrier level. In addition, as PSPI data spanned several disparate databases, there was no facility to drill-down into individual records to better understand identified risks. The aim was to build on the use of existing PSPIs to create more granularity and insight. The approach was to design a data model with PSPIs recorded at the lowest level, i.e. by equipment, and aggregate upwards. This was achieved by mapping equipment to SECEs and QRA Events based on the equipmentâ€™s function, parent system and physical location on the asset. The mapping of equipment to Barriers and MAHs could then be inferred. Crucially, all insights had direct line of sight back to the source data to support decision-making.

For the PoC, our scope included two MAHs and a subset of dynamic1 hardware barriers. Further work is ongoing to scale up the enhanced barrier model and integrate the data with the GDi Vision tool to provide a risk-informed digital walkthrough of the asset.

The three case studies demonstrate some of the additional insights and outputs that are possible to uncover from existing operational data. Joining datasets together and visualising them in an intuitive way with our Proof-of-Concept enhanced barrier model, we were able to give our client, AOC, a more granular and useful view of their MAH, SECE, and QRA data. With our Detect tool, we, in partnership with BP, were able to model the relationship of different datasets with process safety events, enabling users to focus in on the Performance Influencing Factors that were most relevant in a process safety context. Finally, we leveraged the power of large language models to enhance the quality and insight gained from observations. Data science offers a wide range of techniques, from data mapping all the way to advanced AI models. We apply the right methodology and technique to the right data and use case to open the door to a new level of insight. While doing this, we maintain direct line-of-sight to the ultimate purpose of our projects and product development: to ensure the safety of the industry and every person working in it.

STORM
Safety Tool for Operators and Remote Monitoring in Early Leak Detection

Stéphane Collas, Xavier Watremez
TotalEnergies

Summary

1. Objectives
2. TADI - TotalEnergies Anomaly Detection Initiatives
3. Safety Monitoring
4. STORM – Safety Tool for Operators & Remote Monitoring
5. Industrial pilot

Visuals and data from tests carried out by TotalEnergies on its TADI platform
© All rights reserved – 2024

#1

Objectives

Preventing technological risks

Visuals and data from tests carried out by TotalEnergies on its TADI platform
© All rights reserved – 2024

Objectives : preventing technological risks

Need for improvement

- Lack of detection performance
- Leak source location needs post-analysis
- Anticipation (early detection) required to avoid escalation

Industry 4.0

- Technological acceleration for Safety
- Improving the prevention using digitalization and twins
- Combining innovative technologies & data

Operator of the future

- Key player (control, responsability & adaptation)
- Situation awareness without mental load for decision making
- Less exposed to risks

Visuals and data from tests carried out by TotalEnergies on its TADI platform
© All rights reserved – 2024

#2

TotalEnergies

TADI

TotalEnergies Anomaly Detection Initiatives

Test infrastructure dedicated to safety, to environment and robotic developments

Visuals and data from tests carried out by TotalEnergies on its TADI platform
© All rights reserved – 2024

TADI - TotalEnergies Anomaly Detection Initiatives in Lacq (64) – France

TotalEnergies

TADI is a unique test infrastructure in Europe, dedicated to safety (major accident prevention), to environment (GHG reduction) & robotic developments. It is an operational theater, allowing assessing technologies for the detection & quantification of gas emissions (CH_4, CO_2, N_2, etc.).

Visuals and data from tests carried out by TotalEnergies on its TADI platform
© All rights reserved – 2024

TADI General description

TotalEnergies

- What is TADI?
  - 2000 m² (40 m x 50 m) on PPL site, equipped with real industrial equipment
  - Currently gas emission: CH_4, CO_2, N_2 (with bottle and rack)
  - Currently flow range: 0,001 – 300 g/s
  - Gas interferers: C_2H_6, C_3H_8…
  - Multi sources gas emission
  - Dedicated control room, with logic controller, for process regulation and management
  - Tunable installation for specific needs on request
  - About 50 technologies evaluated in which several mobile technologies (handcam, drone, aircraft, satellite, robot…)
  - Currently on site 30 fixed sensors hosted (24/24), acoustic, infrared, weather station…
  - Collaboration with METEC (Methane Emissions Technology Evaluation Center) Colorado State University

Visuals and data from tests carried out by TotalEnergies on its TADI platform
© All rights reserved – 2024

#3

TotalEnergies

Safety Monitoring

Innovative and complementary technologies for early gas leak detection

Visuals and data from tests carried out by TotalEnergies on its TADI platform
© All rights reserved – 2024

Mobile solutions of early gas leak detection Technologies benchmark on TADI

TotalEnergies

OGI Optical Gas Imaging	ALI Acoustic Leak Imaging
Specific to a range of gas	Valid for all gases
Plume visualization	Clear indication of leaks *(easy image interpretation)*
Sensitive to wind and humidity	Sensitive to ultrasound noise
Post-processed quantification	Real time quantification
Calibration recommended	No calibration needed
Expertise level: high	Expertise level: moderate
Price range: high	Price range: medium

Visuals and data from tests carried out by TotalEnergies on its TADI platform
© All rights reserved – 2024

Fixed solutions of early gas leak detection Technologies benchmark on TADI

TotalEnergies

Development and qualification of gas leaks detection

Optical & acoustic technologies

Tests & permanent monitoring in TADI

Operational validation on :

- Conventional oil & gas sites
- H_2 production and delivery sites

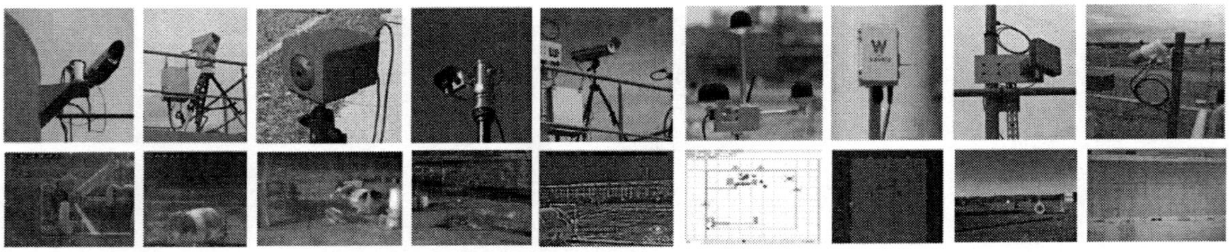

Examples of optical and acoustic technologies tested on TADI

Visuals and data from tests carried out by TotalEnergies on its TADI platform
© All rights reserved – 2024

Fixed solutions of early gas leak detection Technologies benchmark on TADI

GIS Gas Imaging System	AGLED Acoustic Gas Leak Early Detection
Specific to a range of gas	Valid for all gases
Plume visualization	Localization of leak source
Sensitive to wind and humidity	Sensitive to high frequency noise
Real time detection & classification	
Maintenance needed: low	
Expertise level: low	
Price range: high	Price range: medium

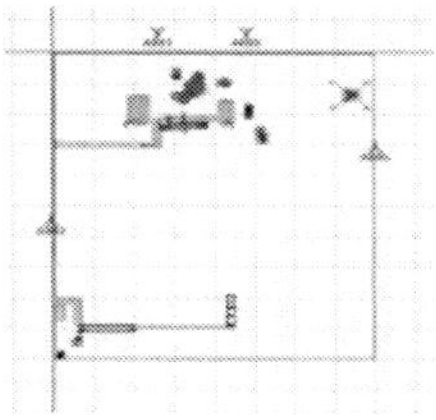

Visuals and data from tests carried out by TotalEnergies on its TADI platform
© All rights reserved – 2024

#4

STORM

Safety Tool for Operators & Remote Monitoring

Decision-aid tool capable of cross-referencing relevant safety data from different sources to generate immediate high-level synthetic risk assessment

Digitalisation to support safety ambitions

Visuals and data from tests carried out by TotalEnergies on its TADI platform
© All rights reserved – 2024

Global architecture

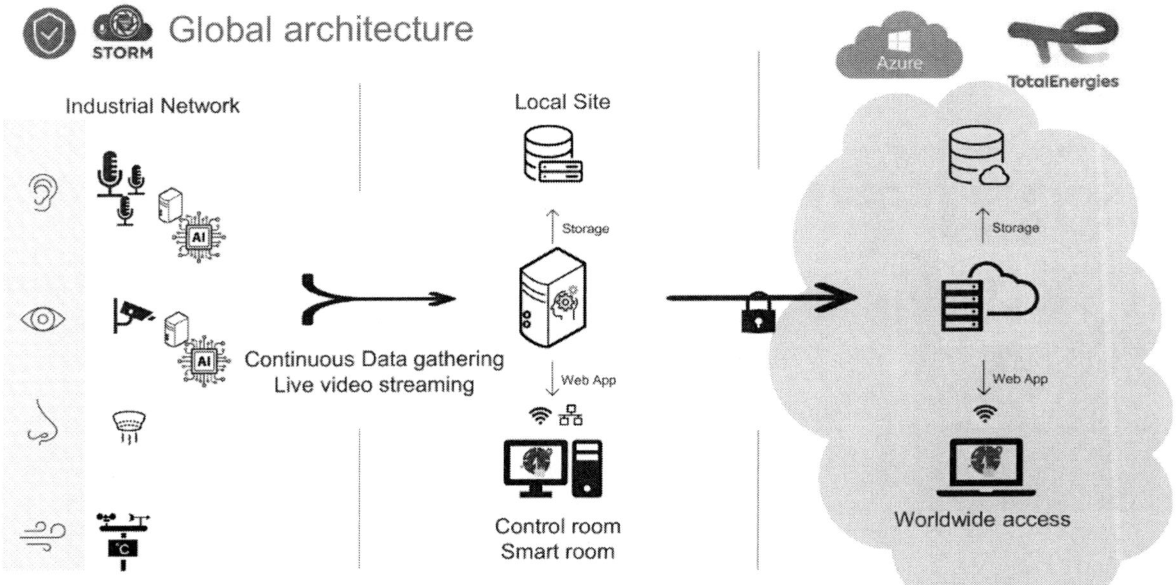

Visuals and data from tests carried out by TotalEnergies on its TADI platform
© All rights reserved – 2024

Application in TADI

➡ Fully controlled replay with synchronized alarms - data - video

➡ Easy & valuable analysis of all recorded data

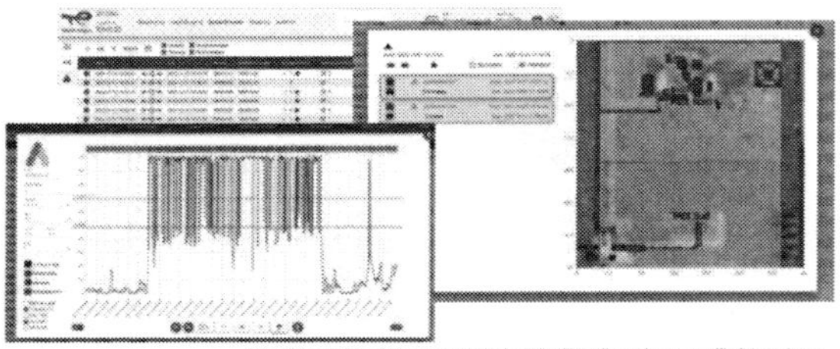

➡ Geographic Information System with live monitoring of:

- Sensor status
- Warning or alarm with location
- Multi cam streams
- Weather, concentration, sensor data…
- Timeline of events
- Acknowledgment system

Visuals and data from tests carried out by TotalEnergies on its TADI platform
© All rights reserved – 2024

#5

TotalEnergies

Industrial pilot

24/7 monitoring of compression zones

STORM

Visuals and data from tests carried out by TotalEnergies on its TADI platform
© All rights reserved – 2024

AGLED and GIS implementation

24 ATEX AGLED antennas & 3 Teledyne FLIR G300a IR cameras & IMC Service IR-SYM-VOC monitoring system, cables, fiber optics, servers installed in compressions zones since 2023

IMC Service TotalEnergies

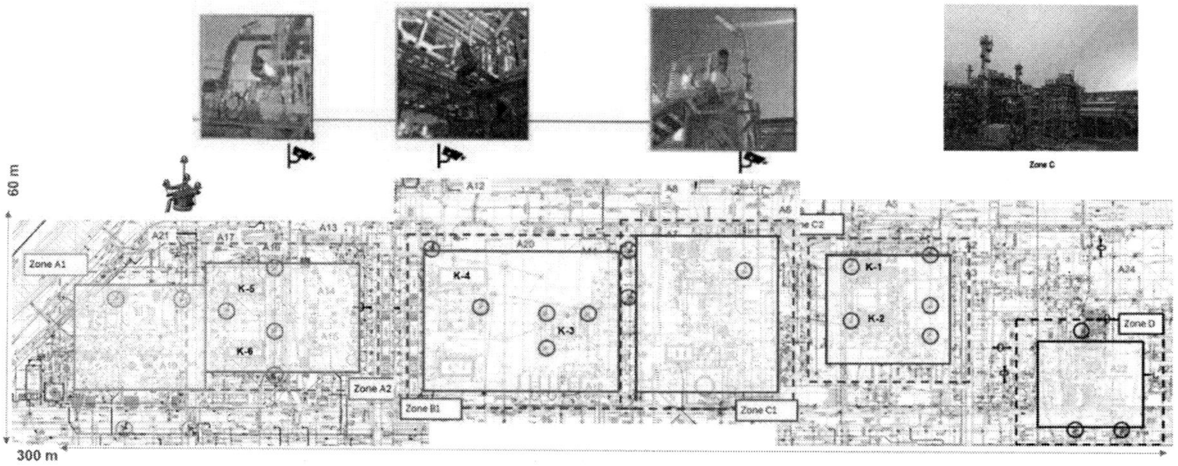

Visuals and data from tests carried out by TotalEnergies on its TADI platform
© All rights reserved – 2024

24/7 monitoring

STORM

Rotating machines monitoring, higher leak risk

TotalEnergies

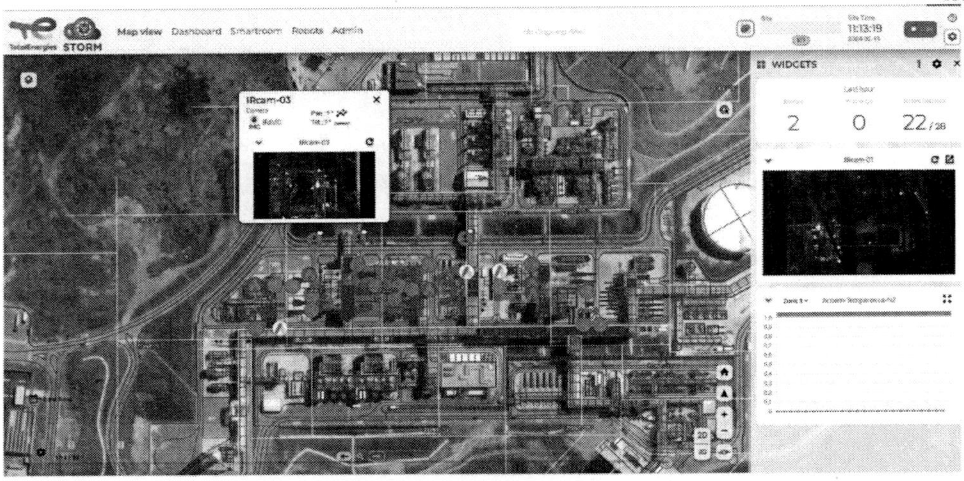

Visuals and data from tests carried out by TotalEnergies on its TADI platform
© All rights reserved – 2024

Conclusion

TotalEnergies

- Complementary technologies for safety improvement
- Homemade solution to ensure cybersecurity compliance and diversity of input technologies
- Convergence of safety and environmental monitoring

Visuals and data from tests carried out by TotalEnergies on its TADI platform
© All rights reserved – 2024

TotalEnergies

Thank you

Visuals and data from tests carried out by TotalEnergies on its TADI platform
© All rights reserved – 2024

Avertissement - Propriété intellectuelle

TotalEnergies

Définition TotalEnergies / Compagnie
Les entités dans lesquelles TotalEnergies SE détient directement ou indirectement une participation sont des personnes morales distinctes et autonomes. Les termes « TotalEnergies », « compagnie TotalEnergies » et « Compagnie » qui figurent dans ce document sont utilisés pour désigner TotalEnergies SE et les entités comprises dans le périmètre de consolidation. De même, les termes « nous », « nos », « notre » peuvent également être utilisés pour faire référence à ces entités ou à leurs collaborateurs. Il ne peut être déduit de la simple utilisation de ces expressions une quelconque implication de TotalEnergies SE ni d'aucune de ses filiales dans les affaires ou la gestion d'une autre société de la compagnie TotalEnergies.

Avertissement
Cette présentation peut contenir des déclarations prospectives, au sens du Private Securities Litigation Reform Act de 1995, relatives à la situation financière, aux résultats d'exploitation, aux activités, à la stratégie et aux projets de TotalEnergies, qui sont soumis à des facteurs de risque et à des incertitudes résultant de changements dans, notamment, le développement et l'innovation technologiques, les sources d'approvisionnement, le cadre juridique, les conditions de marché, les événements politiques ou économiques. TotalEnergies n'assume aucune obligation de mettre à jour publiquement les déclarations prospectives, que ce soit en raison de nouvelles informations, d'événements futurs ou autres. De plus amples informations sur les facteurs susceptibles d'affecter les résultats financiers de la Compagnie sont fournies dans les documents déposés par TotalEnergies auprès de l'*Autorité des Marchés Financiers* et de la US Securities and Exchange Commission. En conséquence, aucune certitude ne doit être accordée à l'exactitude ou à la justesse de ces déclarations.

Propriété intellectuelle
Toute reproduction, publication, transmission ou plus généralement toute exploitation des éléments de cette présentation est interdite, sauf autorisation écrite expresse de TotalEnergies.

Visuals and data from tests carried out by TotalEnergies on its TADI platform
© All rights reserved – 2024

SIEMENS energy

Adaptive Pressure Relief System Design:

Correlation-based Modeling for Relief Capacity at Early Capital Project Stage

2024 Global Conference on Process Safety & Big Data
October 29-30, 2024

Nancy Faulk and Anushree Datta

Siemens Energy is a registered trademark licensed by Siemens AG

Unrestricted © Siemens Energy, 2024

SIEMENS energy

Nancy Faulk

- CCPS Certified Process Safety Engineer
- 30 years of experience in process engineering, risk assessment, process safety
- Technical Advisor – Siemens Energy Process & Safety Consulting group since 2010
- B.S. and Ph.D. in Chemical Engineering from Texas A&M University
- API 580 (Risk-based Inspection) certified
- Served on CCPS Projects for Process Safety Metrics, Aging Process Facilities, and the Book of Beacons

Anushree Datta

- Master of Science in Chemical Engineering from Louisiana State University
- Joined Siemens Process & Safety Consulting group in 2007
- Process Safety Engineer for 6 years
- Project Engineer for 6 years
- Program Manager for 3 years
- Engineering Manager – Safety Consulting (2 years)

Unrestricted © Siemens Energy, 2024

SIEMENS energy

Adaptive Pressure Relief System Design:

Correlation-based Modeling for Relief Capacity at Early Capital Project Stage

2024 Global Conference on Process Safety & Big Data
October 29-30, 2024

Nancy Faulk and Anushree Datta

Siemens Energy is a registered trademark licensed by Siemens AG.

Unrestricted © Siemens Energy, 2024

Disclaimer

The information contained in this presentation represents the current view of the authors at the time of presentation. Process safety management is complex, and this presentation cannot embody all possible scenarios or solutions related to compliance. This presentation is for informational purposes only. Siemens Energy makes no warranties, express or implied, in this paper.

Unrestricted © Siemens Energy, 2024

Agenda

1. Evaluation of Pressure Relief Requirements
2. Use Case: Crude Tower
3. Correlation-based Modeling and Results
4. Next Steps
5. Closing Remarks

Unrestricted © Siemens Energy, 2024

Evaluation of Pressure Relief Requirements

- Relief device sizing / design basis used to determine size of relief device(s)
- Current practice: Revalidating any CAPEX/OPEX changes on existing pressure relief systems
  - Analyze effects of new feed on physical properties
  - Followed by Column simulation
  - Note, this exercise may take several weeks
- **Challenge: Determine 'one' controlling basis on which to conduct a pressure relief system study**
- Feeds used vary significantly and making changes has an impact to the relief system ***which may be an invisible risk***

Unrestricted © Siemens Energy, 2024

- Idea: Develop models / algorithms to predict pressure relief and flaring requirements
  - *Which are as accurate as current detailed pressure relief studies*

Can Big Data Be Used for Modeling Relief Rates?

- Large data lakes available
  - 30+ years of pressure relief studies
  - Operating history (e.g., DCS, PI system data)
- Is there potential
  - to monitor systems of interest and understand the **real time impact** of feed changes on relief requirements?

Unrestricted © Siemens Energy, 2024

Agenda

1. Evaluation of Pressure Relief Requirements
2. Use Case: Crude Tower
3. Correlation-based Modeling and Results
4. Next Steps
5. Closing Remarks

Unrestricted © Siemens Energy, 2024

Crude Oil Supply Changes

- Price of crude oil / value of refined products: **highly volatile**
- Refineries often vary their crude diet, based on:
  - Product desirability / Availability of different crude blends
- Different blends may have different properties
- **The refinery must be able to transition efficiently and safely among various crude blends and throughput rates**
  - To optimize operation, economics, and consumer market demands

Unrestricted © Siemens Energy, 2024

Crude Unit Feed

- Usually a mix of crude oils from tankage
- **Vary widely in physical characteristics**
- Crude oils are complex mixtures
  - Consistency from water-like liquid to tar-like solids
  - Color from clear to black
  - Contains water, salts, solids, and trace metals
  - Crudes referred to as "sweet" or "sour" based on weight percent of sulfur

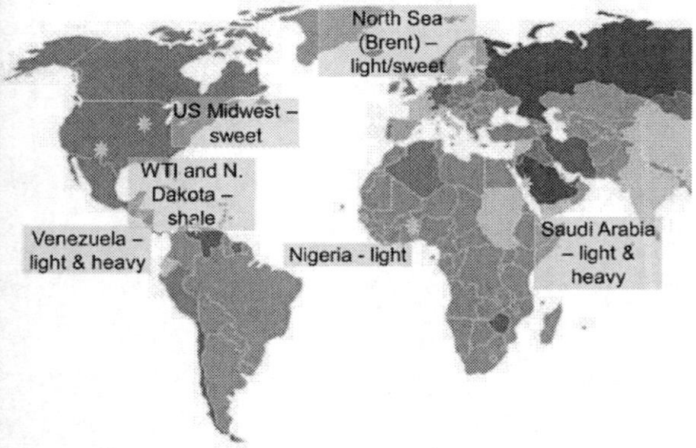

Unrestricted © Siemens Energy, 2024

Crude Refining

Crude Blend ⟶

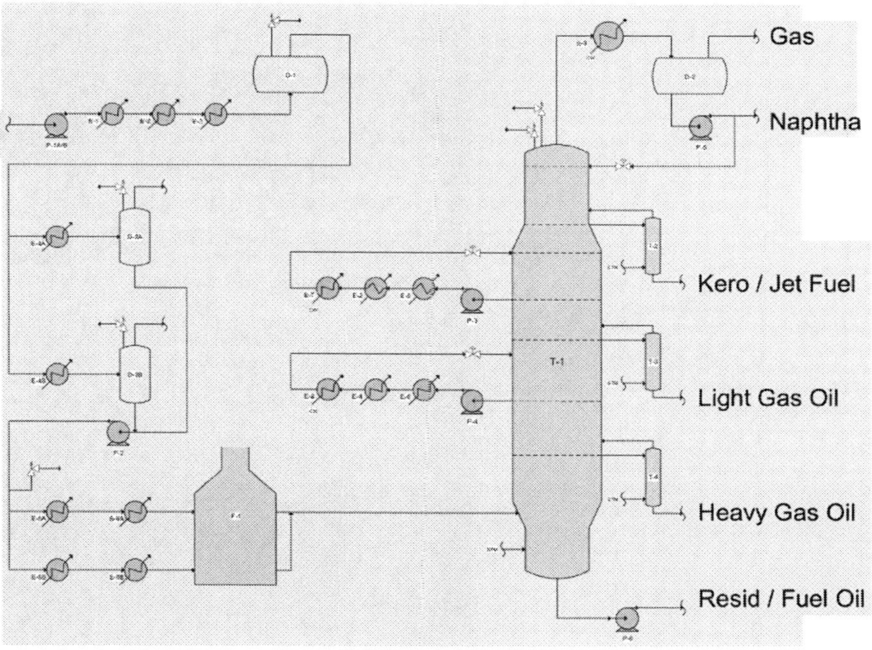

- Crude separated into fractions
- Operates near atmospheric pressure
- Feed enters at 750°F (400°C)
- Overhead temp ~300°F (~150°C)
- Bottoms temp ~1000°F (~540°C)

Problem Statement

"The relief system may be adequate for 100MBPD of a 50/50 blend of crudes A & B, but what would be the answer for a different distribution?

Is there a quick and reliable way to understand the impacts of changing crude slates* without *performing detailed calculations and simulations?"

Siemens Energy Process & Safety Consulting 12

Unrestricted © Siemens Energy, 2024

$$API gravity = \frac{141.5}{SG_{60^\circ F}} - 131.5$$

Crude Oil Characterization

- Petroleum is characterized by bulk density (API gravity), sulfur content, and boiling point curve
  - Crude **quality** is generally measured by its gravity or density and its sulfur content
  - Low API gravity (high density) and high sulfur content depict low quality crude

Siemens Energy Process & Safety Consulting 13
Unrestricted © Siemens Energy, 2024

Agenda

1. **Evaluation of Pressure Relief Requirements**
2. **Use Case: Crude Tower**
3. **Correlation-based Modeling and Results**
4. **Next Steps**
5. **Closing Remarks**

Siemens Energy Process & Safety Consulting 14
Unrestricted © Siemens Energy, 2024

- The goal for the model: which parameter particular to the crude blend may best correlate it to the relief device capacity
  - Perhaps a physical property or operating condition?
- Relief systems database available to provide data sets for model

Modeling Goals

- Can a model provide answers?
  - Risk Visibility: Range of safe operating limits for any Crude Blend combinations
  - Opex Flexibility : Empower operations for Crude Diet Flexibility
  - Capex Solution: Maximum % production for certain products, PRD size requirements
  - Timely Answer: Quick answer on Crude Blend impact on Relief Systems

Unrestricted © Siemens Energy, 2024

Study was performed by:

1. Identify which different blends were being fed to the unit
2. Perform relief simulation for controlling scenario (traditional methodology)
3. Vary the blends in the relief simulation, and repeat simulations
4. Import data from simulation into digital relief systems database, for sizing PRD
5. Estimate % of relief valve capacity used for different crude blends

Model Development

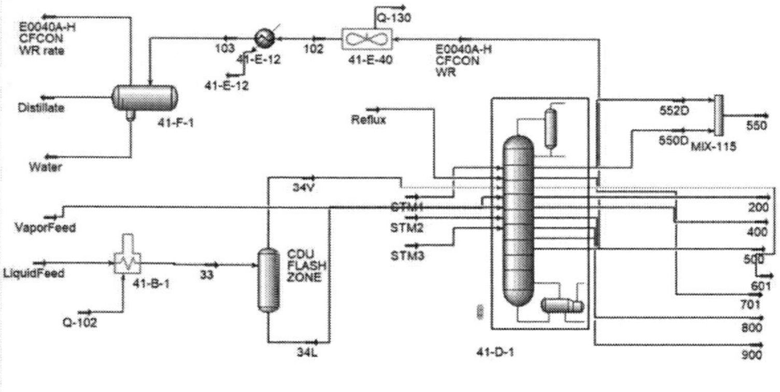

Unrestricted © Siemens Energy, 2024

Selection of Properties for Testing

- Many physical properties of the Crude Blend, as well as some of the operating conditions, can affect the relief rate
- The following data, among others, were exported from the digital database for several crude blend combinations:

- Molecular weight
- Feed rate
- API gravity
- Reid vapor pressure
- Production rate of higher quality products
- Required relief rate for sizing case

Unrestricted © Siemens Energy, 2024

Crude Column Relief Valve Design Basis
Detailed Analysis vs Data-driven Model

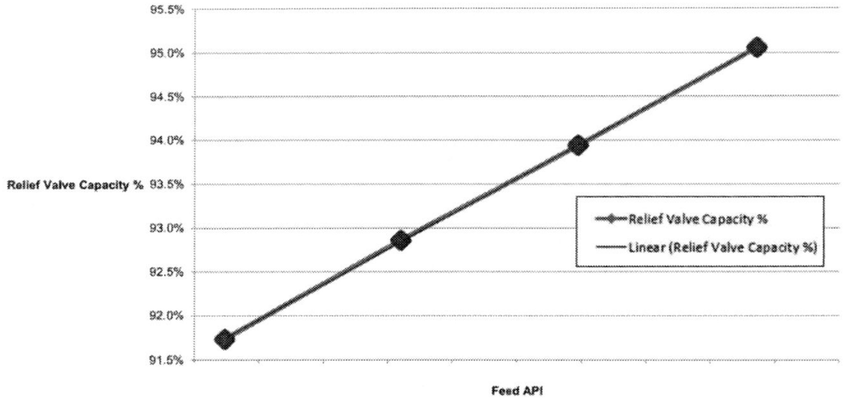

Unrestricted © Siemens Energy, 2024

Relief Capacity versus Estimated Product Rate, Based on Crude Blend Feed

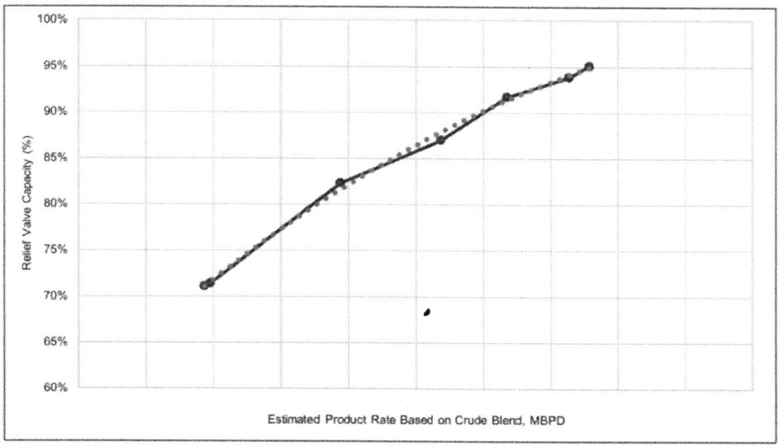

Unrestricted © Siemens Energy, 2024

Agenda

1. Evaluation of Pressure Relief Requirements
2. Use Case: Crude Tower
3. Correlation-based Modeling and Results
4. Next Steps
5. Closing Remarks

Unrestricted © Siemens Energy, 2024

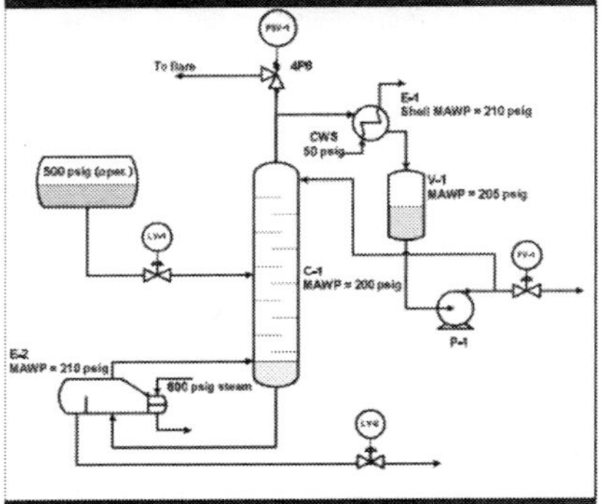

Impact on Systems Other than Crude Columns

- What about the other column systems?
  - Other refinery units?
  - It is expected that they would be affected by a change in the crude slate
- Is it possible that there could there be a correlation specific to columns within other process units? To the flare capacity?
- **Further studies to be performed**

Unrestricted © Siemens Energy, 2024

Agenda

1. **Evaluation of Pressure Relief Requirements**
2. **Use Case: Crude Tower**
3. **Correlation-based Modeling and Results**
4. **Next Steps**
5. **Closing Remarks**

Unrestricted © Siemens Energy, 2024

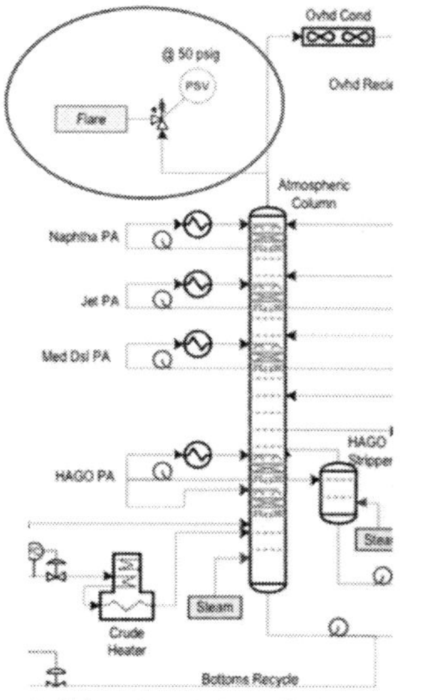

Summary: Required Relief System Capacity
Can It Be Predicted During Conceptual Stage?

Potential to:

- Determine which blends are likely to be the controlling case for relief system sizing without detailed analysis of all blends
- Determine how new blend combinations would impact relief system capacity without performing detailed calculations and sizing for each combination
- Determine what maximum % production is achievable for each product while maintaining adequacy on the relief system, without performing multiple simulations

Unrestricted © Siemens Energy, 2024

Questions?

SIEMENS energy

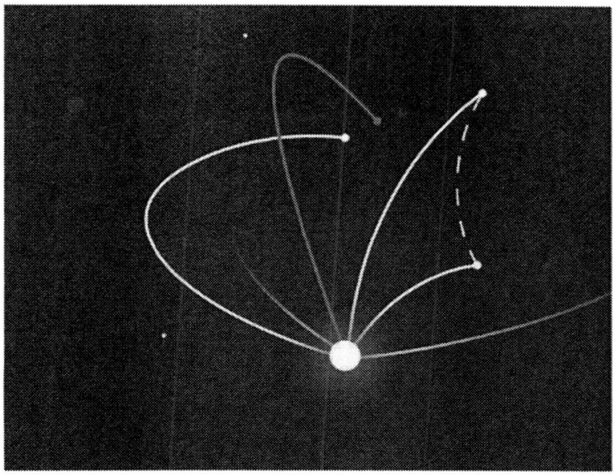

Siemens Energy is a registered trademark licensed by Siemens AG.

Unrestricted © Siemens Energy, 2024

A Review Paper on Process Safety

Conference

Global Conference on Process Safety and Big Data

Year

2024

Proceeding

2024 Global Conference on Process Safety and Big Data

Group

General Program

Session

Business Cases for Digitalization / Big Data Analytics

Time

Tuesday, October 29, 2024 - 5:25pm to 5:50pm

Authors

Edwards, L. - Presenter

This review considers process safety case studies and standards over the last 10 years. The aim of this review is to gather relevant data on what has been encountered in the Process Industry in recent years, in order to guide process safety standards in charting the way forward.

Leveraging Inspection Database Management System to Bridge between Data Analytics and Big Data

Conference

Global Conference on Process Safety and Big Data

Year

2024

Proceeding

2024 Global Conference on Process Safety and Big Data

Group

General Program

Session

Business Cases for Digitalization / Big Data Analytics

Time

Tuesday, October 29, 2024 - 5:50pm to 6:15pm

Authors

Jamshidi, P. - Presenter, Siemens Energy
Biava, M.
Inspection reduces uncertainty by identifying the true damage state. Inspection only sample the overall area of a plant which may lead to inaccurate and unrealistic outcome. In this study, big data is generated with the use of an advanced technique to collect millions of readings coupled with visual images for various forms of degradations. The data is then overplayed on realistic models to be processed by an IDMS with embedded statistical methods to improve risk and reliability analysis for better predictions of future probability of failure. This work provides an overview over this advanced integrity management approach using the big data. Through this integration of big data, IDMS and customized engineering analysis, users will be able to reduce the risk while maximizing asset performance.

Transforming Process Safety with Digital Twin and Augmented Reality: A New Era in Permit-to-Work and LOTO Management

Conference

Global Conference on Process Safety and Big Data

Year

2024

Proceeding

2024 Global Conference on Process Safety and Big Data

Group

General Program

Session

Business Cases for Digitalization / Big Data Analytics

Time

Tuesday, October 29, 2024 - 5:00pm to 5:25pm

Authors

Kunze, S. - Presenter, Hybird Tech
Kanfoud, J. Sr.
In the rapidly evolving landscape of industrial operations, ensuring process safety and efficiency remains paramount. The integration of advanced technologies such as augmented reality (AR) and mixed reality (MR) into permit-to-work (PTW) digitisation marks a transformative leap forward. HyBird exemplifies this advancement by leveraging digital twin capabilities and AR features to enhance safety, streamline workflows, and capture data with perfect context.

This abstract explores the critical role of AR and MR in preventing errors and ensuring data accuracy, alongside the digitisation of PTW and Lockout-Tagout (LOTO) processes. AR and MR technologies facilitate immersive interactions with digital twins, providing real-time, contextual data that enhances decision-making and operational efficiency. The digital twin technology offers a photorealistic 3D visualization of facilities, continuously updated via a mobile app using AI algorithms. This ensures the digital twin remains current, relevant, and accurately contextualized, thereby preventing misunderstandings and reducing reliance on workers' tacit knowledge.

Furthermore, the AR mobile app overlays digital information onto the physical world, enabling workers to visualize equipment data, documents, and maintenance workflows directly in their field of view. This immersive interaction significantly reduces potential errors by providing precise, real-time data about the assets. The app also enables tagging of defects and hazards, which are geotagged and integrated into the digital twin, eliminating the need for additional office reporting.

Digitising the PTW process with HyBird capabilities represents a significant leap in safety and efficiency. Traditional PTW processes are often manual and paper-based, making them difficult to maintain and audit. By incorporating digital twin and AR technologies, HyBird provides a 3D context to PTW and LOTO planning, allowing for more robust and precise execution of these critical safety procedures.
During LOTO planning, detailed visualization ensures that all steps are thoroughly understood before execution, mitigating risks and preventing accidents. The integration of smart P&IDs with the digital twin enables quick identification and location of components, essential for effective isolation and permit management. This is particularly beneficial in congested areas or for onboarding new staff unfamiliar with the facility layout.

During the execution phase, the AR mobile app guides operators through each step of the PTW and LOTO processes. By overlaying digital instructions onto the physical environment, the app ensures workers follow correct procedures, significantly reducing human error. This enhances operational safety and efficiency by providing real-time data on equipment specifications, maintenance history, and safety procedures.

One of the most significant advantages of these technologies is their ability to capture and contextualize data accurately. AR and digital twin technologies ensure all data is spatially and temporally contextualized, transforming tacit knowledge into explicit knowledge accessible to the entire organization. This capability enhances safety, improves maintenance strategies, reduces downtime, and extends equipment life.
The integration of AR and MR into industrial operations, particularly in PTW digitisation, represents a significant advancement in process safety and efficiency. By providing a 3D context to PTW and LOTO planning and execution, these technologies mitigate risks and transform industrial asset management. This abstract highlights the transformative potential of these technologies in ensuring safer, more efficient industrial operations.

Leveraging Operational Risk Management (ORM) Digital Twins for Enhanced Process Safety in Hazardous Industries

Conference

Global Conference on Process Safety and Big Data

Year

2024

Proceeding

2024 Global Conference on Process Safety and Big Data

Group

General Program

Session

Generative Artificial Intelligence and Big Data Analytics for Process Safety

Authors

Bartlett, A. - Presenter, Sphera

Explore how Operational Risk Management (ORM) Digital Twins are reshaping risk management in hazardous industries. These innovative solutions offer real-time insights into the relationship between operational decisions and process safety barriers, enhancing safety standards. Leveraging advanced technologies like IIoT and predictive analytics, ORM Digital Twins empower proactive risk identification and mitigation. Join us to discover the transformative potential of ORM Digital Twins through case studies and expert insights, revolutionizing safety practices in hazardous environments.

The quest for a fundamental upgrade in chemical operator qualification methods

Conference

Global Conference on Process Safety and Big Data

Year

2024

Proceeding

2024 Global Conference on Process Safety and Big Data

Group

General Program

Session

Generative Artificial Intelligence and Big Data Analytics for Process Safety

Authors

Voges, S. - Presenter, Voovio
Janssen, B., Voovio

At last year's presentation we discussed the benedictions of digitally enhancing the chemical operators' learning experience in the process of getting acquainted to a new production plant. Enhanced Reality technology, described in (1), (2), (3), is the only solution to offer simulation capabilities and photorealism at the same time and therefore considered as the state of the art to achieve this.

Focus was on avoiding human error in the context of process safety as a primary motivation. While this remains highly relevant, we will now provide a broader perspective and share our experience on what drives key stakeholders in chemical operations to start looking into VR, AR, XR solutions for training. We will present several examples for real use cases in chemical and petrochemical industry and how these were translated into highly viable business cases.

Still, why do so many digital initiatives fail? Achieving the full potential of a project involves proper preparation and smooth project execution. However, this is not enough: the adoption process deserves special attention as well - as any other change process it may come with pitfalls. We share insight on the key success factors for successful implementation that we found in our practice.

How Dynamic Barrier Management Drives Effective Process Safety Management

Conference

Global Conference on Process Safety and Big Data

Year

2024

Proceeding

2024 Global Conference on Process Safety and Big Data

Group

General Program

Session

Process Safety Management and Permit To Work

Time

Wednesday, October 30, 2024 - 4:40pm to 5:05pm

Authors

Mejias, R. - Presenter, Wolters Kluwer - Enablon
Versteegh, F., Wolters Kluwer - Enablon

In the context of process safety, the dynamic management of barriers and safeguards is critical to ensuring robust process safety. Traditional methods of managing process safety barriers are often static, leading to potential degradation and a lack of real-time visibility into their status. Organizations that rely solely on these outdated methods risk operating with weakened barriers and an incomplete understanding of their control systems' effectiveness.

Dynamic barrier management represents a transformative approach, offering real-time insights and continuous updates on the condition of safety barriers. This proactive management style not only identifies barrier degradation as it occurs but also enables immediate corrective actions, thereby maintaining the health and effectiveness of these critical safeguards. The essence of dynamic barrier management is to ensure that barriers are not just installed but are actively maintained and monitored, guaranteeing their functionality when needed most.

To achieve this level of dynamic management, organizations are increasingly moving away from fragmented Process Safety Management (PSM) software solutions. Instead, they are adopting comprehensive, integrated platforms that unify various aspects of process safety into a single, coherent system. This integration bridges the historically significant gap between engineering and operationsâ€"a critical step towards achieving Operational Excellence. By unifying these functions, organizations can ensure that safety protocols are not only designed effectively but are also implemented and maintained with precision in the operational environment.

Furthermore, there is a substantial opportunity for companies to enhance their Process Hazard Analyses (PHAs) through advanced software solutions. Traditionally, PHAs have been static documents, reviewed periodically and often disconnected from the daily operations of a facility. However, by transitioning to dynamic PHAs, organizations can operationalize these analyses, providing real-time insights and integrating them into the daily workflow. This shift enables a more responsive and informed approach to managing process safety risks, turning PHAs from a retrospective analysis tool into a proactive, operational instrument.

The adoption of a single, integrated software platform for process safety offers several key advantages. First, it provides a centralized repository for all safety-related data, facilitating easier access and better data management. Second, it enhances communication and collaboration across different departments, ensuring that safety information flows seamlessly between engineering, operations, and management. Third, it supports the continuous improvement of safety processes through real-time data analytics and feedback mechanisms, allowing for timely interventions and adjustments.

In conclusion, the future of process safety management lies in the dynamic and integrated management of safety barriers and PHAs. By leveraging advanced software solutions, organizations can transition from static, fragmented approaches to dynamic, holistic systems that offer real-time insights and continuous improvement. This not only enhances the safety and reliability of operations but also drives Operational Excellence by ensuring that safety processes are seamlessly integrated into the fabric of daily operations.

Join us to explore how cutting-edge solutions can revolutionize your approach to process safety management. Learn how to implement integrated process safety systems, embrace dynamic barrier management, and operationalize your PHAs for real-time insightsâ€"all through a single, powerful software platform. Discover the future of process safety and how it can help your organization achieve unparalleled levels of safety and operational performance.

The Impact of Digital Technologies on Operational Safety and Health (OSH) Control of Work (COW)

Conference

Global Conference on Process Safety and Big Data

Year

2024

Proceeding

2024 Global Conference on Process Safety and Big Data

Group

General Program

Session

Process Safety Management and Permit To Work

Time

Wednesday, October 30, 2024 - 4:15pm to 4:45pm

Authors

Bayoumy, M. - Presenter, Sphera

Introduction

The traditional permit-to-work (PTW) systems in operational safety and health (OSH) were initially developed with paper-based methodologies in mind. These systems have been integral to maintaining safe work environments by ensuring that activities, especially those conducted in confined or complex spaces such as storage areas or process plants, are managed properly. With the advent of digital technologies and the Industry 4.0 movement, there has been a paradigm shift towards adopting smart and agile solutions to enhance the effectiveness and efficiency of these systems. This abstract explores the transformation brought about by digital technologies in the context of OSH COW, highlighting the benefits, methodologies, and motivations behind this shift.

Background

The guidelines for permit-to-work systems, as established by EU/UK standards, underscore their importance as a component of a safe system of work. They provide a framework for managing diverse activities that occur simultaneously in close proximity, thereby mitigating risks associated with such operations. Similarly, OSHA's Process Safety Management (PSM) regulations require rigorous inspection and authorization processes before conducting high-risk activities, such as cutting or welding, to ensure safety protocols are strictly followed. These guidelines and regulations were formulated during an era when paper-based systems were the norm. However, today's digital age necessitates a re-evaluation and modernization of these systems to keep pace with technological advancements and industry needs.

Transition to Digital Permit Systems

The digital age, characterized by the proliferation of smartphones, tablets, and other smart devices, has catalyzed the adoption of digital permit systems. These systems are at the forefront of Industry 4.0, which emphasizes the integration of smart technologies to create more efficient and flexible industrial processes. Digital permit systems offer several advantages over their paper-based predecessors. They allow for the linking of various safety-related activities, such as Permits to Work, Isolations Management/Lockout-Tagout (LOTO), and Job Safety Analysis/Job Hazard Analysis (JSA/JHA), into a cohesive and streamlined workflow. This integration is essential for improving productivity and minimizing risks associated with operational activities.

Benefits of Digital Technologies in COW

One of the most significant advantages of digital permit systems is the ability to visualize cumulative risks in near real-time. By integrating various data points from impaired safety management systems and PSM barrier deviations, these systems provide a comprehensive view of ongoing workloads and their impact on operational activities. This real-time visibility enables operational disciplines such as Maintenance, Engineering, and Planning to make informed decisions that enhance safety and performance.

Furthermore, digital systems offer enhanced visibility and control over operations and maintenance activities across multiple facilities. This increased oversight translates into improved productivity and a noticeable decrease in accidents. Streamlined communication is another critical benefit, as digital permits are easier to follow and can be standardized across various practices and policies. This standardization leads to more consistent safety protocols and a reduction in human errors.

Empowering Operational Excellence

The dynamic nature of modern industrial operations requires a flexible and empowered approach to safety management. Digital permit systems allow organizations to manage risks, locations, and time more effectively, thereby achieving process safety operational excellence. By providing an enterprise technology platform, these systems enable management to make better operational decisions that positively impact safety and operational systems within a facility. This empowerment is crucial for navigating the complexities of contemporary industrial environments, where quick and accurate decision-making can significantly influence overall safety and productivity.

Methodology

Implementing digital permit systems involves a systematic approach to identifying and mitigating risks associated with various work activities. The following methodology outlines the steps involved:

1. **List of Work Activities**: Compile a comprehensive list of all work activities conducted within the facility.

1. **Identify Specific Hazards for Each Location**: For each location within the facility, identify the specific hazards that are present.

1. **List Hazards Associated with Each Activity**: For each work activity, list the hazards that arise from performing the task.

1. **Identify Controls for Hazard Mitigation**: For each identified hazard, determine the controls that need to be in place to mitigate the risk. These controls can include physical safeguards, procedural changes, or administrative measures.

This structured approach ensures that all potential hazards are systematically identified and addressed, thereby enhancing overall safety.

Advantages of a Unified View

A significant advantage of digital permit systems is the ability to gain a single, shared view of the operational reality. This unified perspective is critical for several reasons:

- **Comprehensive Risk Management**: By integrating data from various sources, digital permit systems provide a holistic view of risks and their potential impacts. This comprehensive risk management approach ensures that all stakeholders are aware of the current safety status and can take proactive measures to address any issues.

- **Improved Decision-Making**: Access to real-time data and analytics allows management to make informed decisions that enhance safety and operational efficiency. This improved decision-making capability is essential for maintaining high standards of safety and productivity.

- **Enhanced Communication and Collaboration**: A unified view facilitates better communication and collaboration among different departments and teams. By providing a common platform for sharing information, digital permit systems ensure that everyone is on the same page, leading to more coordinated and effective safety measures.

Conclusion

The transition from paper-based to digital permit systems represents a significant advancement in the field of operational safety and health control of work. Digital technologies offer numerous benefits, including enhanced visibility, improved productivity, streamlined communication, and a unified view of operational reality. By leveraging these technologies, organizations can achieve process safety operational excellence and create safer, more efficient work environments. The structured methodology for implementing digital permit systems ensures that all potential hazards are systematically identified and mitigated, further enhancing overall safety. As the industrial landscape continues to evolve, the adoption of digital technologies will be crucial for maintaining high standards of safety and operational performance.

AUTHOR INDEX

AIChE
120 Wall Street, FL 23
New York, NY 10005-4020

ISBN 979-8-3313-1053-0

CURRAN ASSOCIATES INC.
proceedings.com